T0144492

FM-UWB Transceivers for Autonomous Wireless Systems

RIVER PUBLISHERS SERIES IN CIRCUITS AND SYSTEMS

Series Editors

MASSIMO ALIOTO
National University of Singapore
Singapore

KOFI MAKINWA
Delft University of Technology
The Netherlands

DENNIS SYLVESTER
University of Michigan
USA

*****Indexing: All books published in this series are submitted to Thomson Reuters Book Citation Index (BkCI), CrossRef and to Google Scholar*****

The "River Publishers Series in Circuits & Systems" is a series of comprehensive academic and professional books which focus on theory and applications of Circuit and Systems. This includes analog and digital integrated circuits, memory technologies, system-on-chip and processor design. The series also includes books on electronic design automation and design methodology, as well as computer aided design tools.

Books published in the series include research monographs, edited volumes, handbooks and textbooks. The books provide professionals, researchers, educators, and advanced students in the field with an invaluable insight into the latest research and developments.

Topics covered in the series include, but are by no means restricted to the following:

- Analog Integrated Circuits
- Digital Integrated Circuits
- Data Converters
- Processor Architecures
- System-on-Chip
- Memory Design
- Electronic Design Automation

For a list of other books in this series, visit www.riverpublishers.com

FM-UWB Transceivers for Autonomous Wireless Systems

Nitz Saputra
Qualcomm Inc., USA

John R. Long
University of Waterloo, Canada

Published, sold and distributed by:
River Publishers
Alsbjergvej 10
9260 Gistrup
Denmark

River Publishers
Lange Geer 44
2611 PW Delft
The Netherlands

Tel.: +45369953197
www.riverpublishers.com

ISBN: 978-87-93519-16-9 (Hardback)
978-87-93519-15-2 (Ebook)

©2017 River Publishers

All rights reserved. No part of this publication may be reproduced, stored in a retrieval system, or transmitted in any form or by any means, mechanical, photocopying, recording or otherwise, without prior written permission of the publishers.

Contents

Preface

Significant research effort has been devoted to the study and realization of autonomous wireless systems aimed at wireless sensor and personal-area networking, the internet of things, and machine-to-machine communications. Low-power RF integrated circuits, an energy harvester, and a power management circuit are fundamental elements of these systems. All of these components are designed and demonstrated in the form of a compact, low-power frequency-modulated ultrawideband (FM-UWB) transceiver that is described in this book.

The FM-UWB modulation scheme was chosen because it offers advantages in robustness to interference, and a simple hardware implementation. The FM-UWB transceiver prototypes described in this book utilize advanced circuit design techniques and calibration schemes. They are realized in bulk 65-nm and 90-nm complementary metal-oxide-semiconductor (CMOS) technology. The transceiver is powered by a solar-cell antenna (solant), which transmits and receives RF signals while generating power from light falling on the antenna surface. To realize an autonomous RF system, a CMOS power management circuit is also realized to interface the transceiver and solant. Both transceiver and power management prototypes are described thoroughly in this book. Their performance is validated and compared with existing solutions from the recent technical literature.

The material in this book are divided into 3 main parts:

Chapters 1 and 2 give an introduction to autonomous wireless systems and their application. The FM-UWB scheme, its specification, and prior implementations are described. Also, circuit techniques related to low-power CMOS circuit designs are recommended.

Chapters 3 to 5 describe FM-UWB transceiver implementations in CMOS technology. Block-level circuit diagrams are detailed, along with a novel digital calibration scheme. Measurement results validating the innovations developed for these prototypes are also presented.

Chapter 6 introduces practical energy sources, storage, and DC-to-DC converters for an autonomous chip. A CMOS solar power management circuit customized to power an FM-UWB transceiver is described in detail, and its performance is measured and validated. The operating time for an autonomous FM-UWB system powered by the power management IC is also estimated.

Advanced RF circuit design techniques in CMOS for low-power ultra-wideband FM applications are presented in this book. It is aimed at engineers working in wireless communication industries, academic staff, and graduate students engaged in Electrical Engineering and Telecommunication Systems Research. The material contained in this book are the outcome of theoretical and experimental research efforts undertaken across several years at the Delft University of Technology in the Netherlands. Some material in the book has been presented at Institute of Electrical and Electronics Engineers (IEEE) sponsored conferences and published in IEEE Journals. We hope that this book provides additional, valuable contributions to the readers as a source of technical inspiration or as a reference.

This book would not have been possible without the support and assistance of colleagues, friends, and family. The authors are indebted to the technical staff of the Electronic Research Laboratory (ERL) at the Delft University of Technology. We appreciate the administrative and measurement support provided by ERL members Marco Spirito, Atef Akhnoukh, Ali Kaichouhi, Wil Straver, Loek van Schie, and Marion de Vlieger. We would also like to thank Mina Danesh for providing a solant prototype for our experiments. We are also grateful to Yi Zhao, Wanghua Wu, Morteza Alavi, and Marietya Lauw for proofreading and providing comments that have improved the readability of this text. We are also very grateful to Mark de Jongh and Junko Nakajima of River Publishers for their expert handling of the manuscript.

List of Figures

List of Tables

List of Abbreviations

AC	Alternating Current
ADC	Analog-to-Digital Converter
AGC	Automatic Gain Control
AM	Amplitude Modulation
AM1.5	Air Mass coefficient (1.5 atmosphere thickness)
a-Si	Amorphous Silicon
ASK	Amplitude Shift Keying
AWGN	Additive White Gaussian Noise
BAW	Bulk Acoustic Wave
BER	Bit-Error Rate
BFSK	Binary Frequency Shift Keying
BJT	Bipolar Junction Transistor
BPF	Band-Pass Filter
BPSK	Binary Phase Shift Keying
CDMA	Code-Division Multiple Access
CEPT	Conference of Postal and Telecommunication Administrations
CMOS	Complementary Metal Oxide Semiconductor
DAA	Detect and Avoid
DAC	Digital to Analog Converter
dB	Decibel
DC	Direct Current
DCO	Digitally-Controlled Oscillator
DDS	Direct Digital Synthesis
DEMUX	De-Multiplexer
DR	Dynamic Range
DS	Direct Sequence
DUT	Device Under Test
EIRP	Effective Isotropic Radiated Power
EMI	Electro-Magnetic Interference
ESD	Electrostatic Discharge

ESR	Effective Series Resistance
ETSI	European Telecommunications Standards Institute
FCC	Federal Communications Commission
FDMA	Frequency-Division Multiple Access
FFT	Fast Fourier Transform
FH	Frequency Hopping
FH-CDMA	Frequency-Hopped CDMA
FLL	Frequency-Locked Loop
FM	Frequency Modulation
FM-UWB	Frequency Modulation Ultra-wideband
FPGA	Field-Programmable Gate Array
FSK	Frequency Shift Keying
FSM	Finite-State Machine
GaAs	Gallium Arsenide
GM	Transconductance
HBM	Human Body Model
I/O	Input output
IC	Integrated Circuit
ICO	Current(I)-Controlled Oscillator
IEEE	Institute of Electrical and Electronics Engineers
IF	Intermediate Frequency
INL	Integral Non Linearity
IR-UWB	Impulse-Radio Ultrawideband
ISI	Inter-Symbol Interference
ISM	Industrial, Scientific, and Medical
kbps	Kilobit per second
LDO	Low Dropout (Regulator)
LNA	Low-Noise Amplifier
LOS	Line of sight
LPF	Low-Pass Filter
LSB	Least Significant Bit
LUT	Look Up Table
Mbps	Megabit per second
MIM	Metal-Insulator-Metal
MOM	Metal-Oxide-Metal
MOS	Metal Oxide Semiconductor
MOSFET	Metal Oxide Semiconductor Field Effect Transistor
MSB	Most Significant Bit
MUX	Multiplexer

NBI	Narrowband Interference
NF	Noise Figure
NiMH	Nickel Metal Hydride
NMOS	N-type Metal Oxide Semiconductor
OFDM	Orthogonal Frequency-Division Multiplexing
OOK	On-Off Keying
Opamp	Operational Amplifier
OTA	Operational Transconductance Amplifier
PA	Power Amplifier
PAE	Power-Added Efficiency
PCB	Printed-Circuit Board
PL	Path Loss
PLL	Phase-Locked Loop
PM	Phase Modulation
PMAN	Power Management
PMOS	P-type Metal Oxide Semiconductor
PPM	Pulse Position Modulation
PRBS	Pseudo-Random Binary Sequence
PSD	Power Spectral Density
PSK	Phase Shift Keying
PSRR	Power-Supply Rejection Ratio
PTAT	Proportional to Absolute Temperature
PV	Photo Voltaic
PVT	Process, Voltage, Temperature
Q	Quality factor
QAM	Quadrature Amplitude Modulation
RF	Radio Frequency
RMS	Root Mean Square
RX	Recciver
SAR	Successive Approximation Register
SAW	Surface Acoustic Wave
SC	Switched-Capacitor
SC-CP	Switched-Capacitor Charge Pump
SC-DC	Switched-Capacitor DC-to-DC converter
SIR	Signal-to Interference Ratio
SMA	Sub-Miniature A
SNR	Signal-to-Noise Ratio
Solant	Solarcell Antenna
SOLT	Short Open Load Through

TDMA	Time-Division Multiple Access
TSPC	True, Single-Phase Clock
TX	Transmitter
UGBW	Unity Gain Bandwidth
UWB	Ultrawideband
VCO	Voltage-Controlled Oscillator
VGA	Variable-Gain Amplifier
WBAN	Wireless Body Area Network
WMBAN	Wireless Medical Body Area Network
WMTS	Wireless Medical Telemetry Service
WPAN	Wireless Personal Area Network
WSN	Wireless Sensor Network

1

Introduction

Human is social beings who like to communicate with one another, either actively or passively. In active communication, two or more parties exchange information in a two-way fashion, while in passive observation one party receives information sent by another. Technology provides a medium for information flows that enables such connectivity. An example of active communication medium is the telephone (both wired and wireless); a passive example is a television, and the internet is both passive and active. One element that is common to all of these technologies is a network that allows information or data to be exchanged or spread.

The network can be established using a cable or a wireless channel. Wireless technology requires no physical cable installation and less infrastructure, which is attractive for aesthetic and practical purposes. A wireless connection is advantageous in term of its flexibility and mobility. For example, when the sender and receiver change position, there is no cable that limits their movements or position. On another hand, there are disadvantages to wireless, such as: slower data transmission rate, security risks, and sensitivity to interference. However, the demand for wireless interconnectivity continues to increase, and thus research on wireless communication methods has also flourished [1].

Recently, "the internet of things", whereby sensor-enabled physical objects can communicate with each other is under intensive development [2]. These devices are able to sense parameters from their surrounding environment, and then automatically communicate and coordinate a response to various conditions. Moreover, a better understanding of the environment locally and globally can be obtained by collecting data from many devices. The interconnection between devices on a grand scale could therefore open up new applications such as precision farming, where data collected from wireless sensors in the ground enable crop conditions to be adjusted individually, e.g., by spreading extra fertilizer on areas that need more nutrients [3].

An autonomous wireless system could realize such a network. Autonomous implies that the device is independent of physical infrastructure, and draws so little net energy that it can be powered by a small battery, and/or energy harvested from the local environment (e.g., solar cell). Independence is important because the devices are deployed in a dynamic environment, where they could be self-configured, adapting, and communicating wirelessly. Autonomous wireless devices could be installed in hard to reach places (e.g., underground, to monitors soil, or embedded in a bridge to monitor stress or strain), and must operate maintenance-free for years.

With a limited power source, the output power of the transmitter is kept low to conserve energy, while the receiver sensitivity is limited by the energy available for signal amplification. Thus, the link span between transmitter and receiver is shorter than for a conventional radio, and is only suitable for local connectivity. At a longer distance, data can be relayed across transceivers forming a multi-hop network.

Autonomous wireless devices are constrained at present by their weight, volume and cost of the energy source (e.g., Li-ion battery). Therefore, energy efficiency is of paramount concern for the wireless link. Harvesting renewable energy from the surroundings reduces the battery capacity required and prolongs the operating time, when the transceiver is capable of sending/receiving data and the system is consuming maximum power. Embedded power management is therefore an essential part of the autonomous wireless system to conserve energy. The power management controller also monitors if energy can be harvested, estimates the amount of energy left in storage, and activates the device based on priority. Overall, the power management sub-system ensures that the wireless device functions for as long as possible.

There are many other possible applications for networks of autonomous devices, as illustrated in Figure 1.1. One important application is wireless sensor networks (WSN). Sensors could be deployed to monitor defects due to aging, or provide maintenance alerts in a plant, structure, or building. Sensors monitored via a wireless network avoid tedious and expensive inspections to collect data, and allows the early detection of faults that improves safety [4].

Another application is wireless body-area networks (WBANs) or wireless personal-area networks (WPAN). These networks involve sensing our personal surrounding and provide connectivity between handheld devices and our environment. By automatically adapting to environmental conditions (e.g., turning on a heater when the temperature drops, or turning on music based on our mood as detected by a camera), a WPAN could improve our quality of life [5]. On a bigger scale, a smart home system could be built, enabling

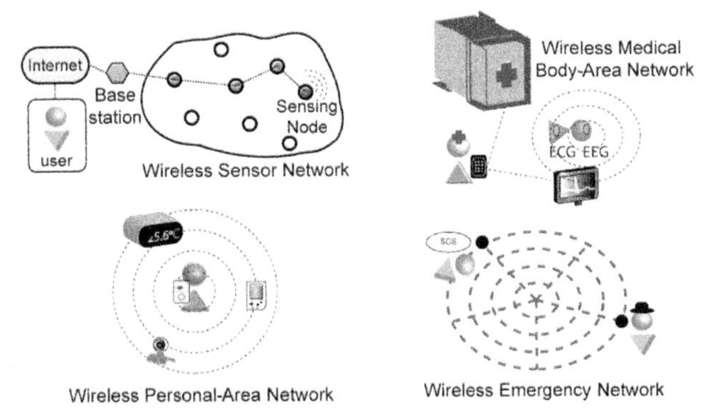

Figure 1.1 Applications for autonomous wireless networks.

automatic management of things such as energy consumption according to demand, monitoring of ambient conditions (e.g., temperature, lighting, music, etc.), and tracking of family members or their pets [6]. On an industrial scale, a smart building integrates building automation, telecommunication, facility management, and energy systems [7].

Wireless medical body-area networks (WMBAN) or wireless medical telemetry services (WMTS) facilitate continuous data streaming between patients being monitored and healthcare practitioners [8]. Monitoring devices should be compact enough so that they can be worn easily without discomfort. The network could monitor one's blood pressure or heart rate at home, and then transmit the data to a medical practitioner via a smartphone and/or personal computer. At a hospital, the network could monitor intensive-care patients' conditions (e.g., respiratory frequency, blood pressure, etc. [9]) thereby improving the chances of survival for critically ill patients. The wireless network also enables long-term monitoring of chronically ill patients, the handicapped or the elderly in real time. This promises to reduce the number of hospital visits and the work load of medical workers and healthcare professionals. Another application in the medical area is bio-metric sensors, such as a non-invasive heart rate monitor using radar [10].

Natural disasters, such as earthquakes or forest fires, require an emergency communication network [11]. A robust and highly redundant ad-hoc network formed by the interconnection of autonomous wireless nodes could provide backup connectivity during a catastrophic event when infrastructure is damaged. The potential applications for the autonomous wireless system are not limited to the ones described above. Other applications and opportunities are likely to emerge when autonomous short distance wireless transceiver devices are widely available and affordable [12].

1.1 Motivation

Autonomous wireless systems have the potential to improve the quality of human life as described in the previous section. The aforementioned applications motivate the development of wireless transceiver hardware for such systems. Though the hardware is our focus in this book, development of the physical layer, network protocols, and standardization are also needed. As more applications emerge and the technology develops, the performance of the system will improve while costs are reduced due to the increase in production volume and higher manufacturing efficiency. This trend will make the devices required to implement the system ubiquitous at an affordable price.

Autonomous wireless systems will likely be deployed on a large scale in the future. Figure 1.2 shows growth in WSN product shipment worldwide (almost doubling every year). Market forecasts predict that tens of billion modules could be shipped over the next decade [14]. These devices would collectively require a lot of energy; assuming that each device consumes just an average of 1 mW, the total power consumption (assuming 10 billion devices) would be on the order of tens of megawatts. Therefore, energy should come from renewable sources collected through scavenging to be sustainable. Energy can be scavenged from sources such as mechanical vibration, light, heat, and microwave radiation [15].

Figure 1.3 shows an example of an integrated autonomous wireless system that might be implemented and deployed in the near future. It consists of a wireless transceiver, baseband circuits, embedded memory, a microcontroller, power management unit, and CMOS sensors. A solar cell to harvest energy is integrated together with the antenna for wireless transmission, while a

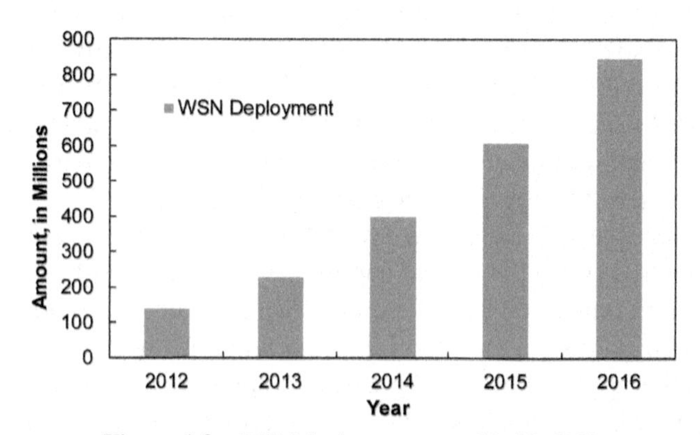

Figure 1.2 WSN deployments worldwide [13].

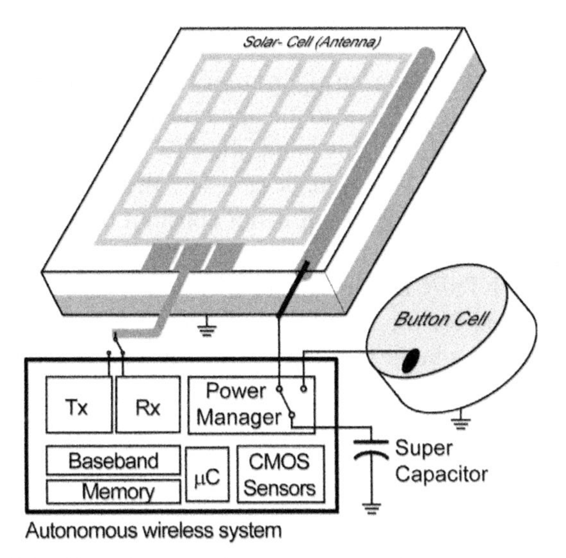

Figure 1.3 Block diagram of an autonomous wireless system example.

button-cell battery is the back-up power source. Harvested energy is stored temporarily on a supercapacitor or by recharging the battery. This device could operate for an indefinite period of time by managing energy harvesting and consumption itself while transmitting information collected by the CMOS sensors to other wireless devices.

Complementary metal-oxide semiconductor (CMOS) is a silicon integrated circuit (IC) technology that can integrate more than 10 billion transistors on-chip (at current time of writing) to realize high performance analog, digital and RF circuits on the same die [16, 17]. System integration reduces the number of external components and IC chips (e.g., a one-chip system). The advantages of integration using CMOS are lower cost in high-volume production, small size, less input/output interfacing, and lower leakage or standby power consumption. Furthermore, the trend in the scaling of CMOS transistors to even smaller dimensions will push integration levels higher in the future which improves RF performance and lowers power consumption [18].

Wireless interfaces often utilize the unlicensed (e.g., ISM) bands currently occupied by WiFi, Bluetooth and Zigbee appliances. However, the potential number of short-range devices operating in a body-area or wireless sensor network would crowd the same space, and the resulting congestion degrades the quality of service and link availability. However, other frequency bands could be used. The US Federal Communications Commission (FCC) [19] authorized the unlicensed use of ultra-wideband (UWB) transceivers in the 3.1–10.6 GHz range, creating an opportunity for UWB transceivers in autonomous wireless systems. Due to the potential for interference with other communication channels in this band, the FCC specified a maximum power emission density for UWB signals of −41.3 dBm/MHz between 3.1 and 10.6 GHz. The spectral density limit is not constant over the entire band as seen from the specific indoor mask proposed by the FCC is shown in Figure 1.4. In Europe, ETSI/CEPT proposed a similar mask [20]. However, the 3.1–5 GHz band used for the transceiver designed in this book needs to employ a detect and avoid scheme in Europe, otherwise the transmitted power spectral density is limited to −70 dBm/MHz [21]. It should be noted that the transceiver operating frequency can be scaled easily. The effective isotropic radiated power (EIRP) spectral density limit and coexistence between UWB and other wireless standards are the subject of ongoing study [22].

The work in this book focuses on the design of a low-power FM-UWB transceiver for short-distance, low-data-rate indoor applications. Simplicity in the transceiver architecture, robustness to interference, and low-cost access to unlicensed spectrum available worldwide motivates this work on a low-power FM-UWB transceiver for autonomous wireless systems. Frequency-modulated ultra-wideband (FM-UWB) is a low-complexity scheme for wireless data communication at rates up to 250 kbit/s across spans less than 10 m [23]. The FM-UWB radio does not require a local oscillator or carrier

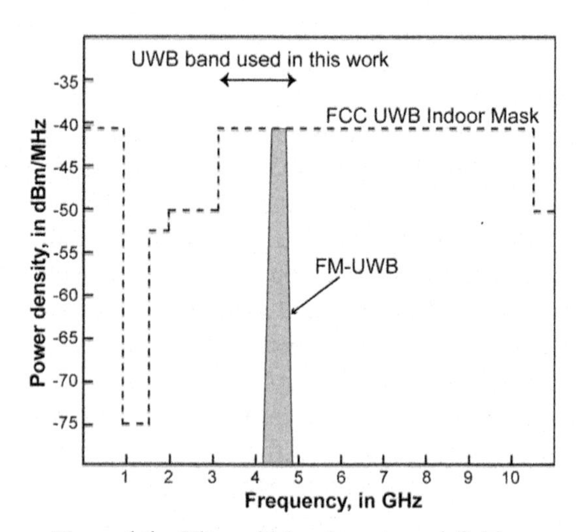

Figure 1.4 Ultra-wideband spectrum definition.

synchronization at the receiver, which results in a simple transceiver architecture that consumes little power in continuous operation [24]. FM-UWB is intrinsically robust against multipath fading and narrowband interference, while mitigation techniques (e.g., using a notch filter or applying detect-and-avoid schemes) may be added to enhance its robustness [25]. An FM-UWB physical layer for body-area networking has also been developed [26]. FM-UWB was recently adopted into the IEEE standard by IEEE802.15 task group 6 [27]. Short-distance transmission allows for a working link at a transmit power less than -10 dBm, and also minimizes interference to other wireless systems nearby. Relatively low transmit power reduces the risk of radiation harmful to the human body, which is important in WBAN and WMBAN applications [5]. A uniform spectral density and steep spectral roll-off at a typical bandwidth of 500 MHz (see Figure 1.4) for the transmit signal power spectrum is easily realized. Co-existence with other communication systems is accommodated by transmitting the signal within a selected portion of the bandwidth allocated for UWB systems, according to the cognitive radio paradigm [28]. The FM-UWB scheme is analyzed and described in more detail in Chapter 2 of this book.

1.2 Design Challenges and Overview

Despite the advantages of implementing an autonomous wireless system in CMOS, the technology introduces performance compromises. For example, the transit frequency (f_T) of a transistor may be lower in CMOS than in bipolar-junction transistor (BJT) technology. Although there are disadvantages to CMOS compared with the BJT, there are also incentives to use CMOS, e.g., seamless integration with digital circuits and manufacturing capacity.

In general, integrating a wireless system on a single chip adds extra complexity and challenges. A circuit has to satisfy the desired specifications regardless of the manufacturing process, supply voltage, or temperature (i.e., PVT) variations to increase its yield in production [33]. Robustness against interference and coexistence with other systems is also necessary for a wireless application. In a wireless transmitter, unwanted interference with other devices is mitigated by limiting the output power, filtering, and accurate control of center frequency or bandwidth of the transmitted signal. A detect-and-avoid scheme can be employed to combat interference at the receiver [34]. This implies that the system must be reconfigurable, and on-chip calibration circuitry must be included to handle PVT variations and interference. Electrostatic discharge (ESD) protection must also be included.

The ESD event could damage the input or output (I/O) interface to the chip, or in the worst case, damage the entire IC. ESD protection in the RF signal path adds parasitic capacitance which limits the bandwidth and could cause signal degradation and distortion [35].

Improving energy efficiency, defined as the ratio of power consumption (in Watts) to data rate (in bit/s), is the key figure of merit for this work. If a continuous operating lifetime of 50 days using a 10-g Li-ion battery (energy density of 120 W-h/kg [36]) is desired, the transceiver should have an energy efficiency better than 10 nJ/bit when operating at a data rate of 100 kbit/s. Conventional radio transceivers (e.g., WiFi, Bluetooth and Zigbee) consume approximately an order of magnitude more energy per bit transmitted, as shown in Figure 1.5 [37–39]. This constrains their operating lifetime, battery size, and potential applications in autonomous systems.

Duty-cycling of a conventional wireless transceiver (i.e., to scale power consumption) may not be possible if real-time data connectivity and high reliability are required. For example, many wireless health-monitoring devices require continuous data streaming between the patient being monitored and healthcare practitioners [40]. Health-monitoring devices use sensors that typically produce continuous streams of data in the range of 1–100 kbit/s, as opposed to multimedia applications that might stream video at 100 Mbit/s [41]. The FM-UWB scheme is suitable for lower data rate applications (see Figure 1.5), although improvement in energy efficiency is required to compete with other schemes (e.g., a wake-up receiver, or super-regenerative transceiver) in autonomous applications such as implantable biosensors.

Some narrowband and ultra-wideband radio technologies have demonstrated energy efficiency suitable for a wireless autonomous system. On the transmitter side (see Table 1.1), power efficiency (i.e., transmitted RF output power divided by the DC consumption) indicates the level of autonomy that may be achieved. On-off keying (OOK) modulation has been a popular

Table 1.1 Transmitter performance summary

Ref.	Year	CMOS Technology (nm)	Modulation	Max. Tx Power (dBm)	DC Power (mW)	Tx Efficiency (%)	Data Rate (Mbit/s)	Energy per Bit (nJ/bit)
[42]	2007	180	OOK	–11.4	3.8	1.9	1	3.8
[43]	2005	180	OOK	–4.4	1.6	22	0.005	320
[44]	2008	90	UWB	–16.4	4.36	0.52	15.6	0.28
[45]	2006	130	BFSK	–5	1.12	28.2	0.3	2.3
[46]	2010	180	BFSK	–5.2	1.15	26.2	0.125	9.2

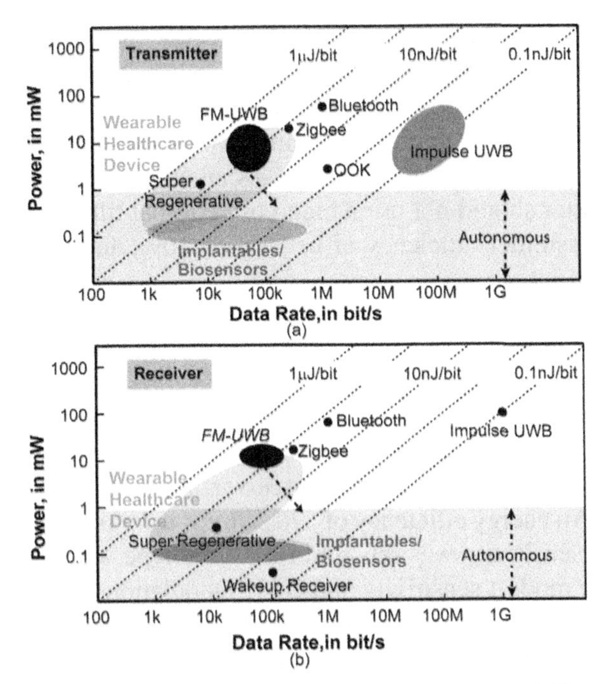

Figure 1.5 Energy efficiency for various wireless (a) transmitters and (b) receivers [37–51].

choice for implementation of ultra-low power radio demonstrators. An OOK transmitter operating with an efficiency of 3.8 nJ/bit at 1 Mbit/s was reported in [42]. The super-regenerative transmitter [43] also employs OOK running at 5 kbit/s, and achieves 320-nJ/bit efficiency and 22% transmitter efficiency. Although similar in total power consumption, the energy efficiencies of these examples are radically different due to the factor 10^3 in their respective data rates. An external high-Q SAW or BAW RF bandpass filter shapes the transmit output spectrum, but adds to the cost of assembly, packaging, and testing. Time-domain impulse radio is a UWB technology that also offers the potential for improved energy efficiency at low data rates. However, the spectral side lobes emitted by an impulse UWB transmitter could interfere with other users. Also, the poor peak-to-average transmit power ratio at low data rates affects the efficiency of the antenna driver adversely [44]. Binary frequency shift keying (BFSK) transmitters ([45] and [46]) achieve transmitter efficiency of 28% and 26%, respectively. A BFSK transmitter benefits from higher efficiency in the power amplifier (PA) that can be designed for the constant envelope BFSK signal. The transmitter in [45] uses a non-standard

0.4 V supply. The transmitter in [46] directly couples the oscillator to the load through a matching network, which causes vulnerability to frequency pulling from strong interference. In general, transmitter efficiency is calculated at maximum output power, and the transmitter which has higher maximum output power usually has a better efficiency. As an attractive alternative, an FM-UWB transmitter should not use a high-cost external filter, use standard 1V supply, has transmitter efficiency of better than 10%, and has an energy efficiency of few nJ/bit.

On the receiver side (see Table 1.2), energy per bit indicates the receiver efficiency. Signal-to-interference ratio (SIR), measured when the bit-error rate (BER) deteriorates to 10^{-3} due to narrowband interference in-band or at a small frequency offset (e.g., 10 MHz), indicates the robustness of the receiver. The narrowband 'wake-up' receiver monitors the wireless channel continuously and activates the main transceiver when addressed by the wireless network. An energy efficiency of 520 pJ/bit at a data rate of 100 kbit/s and −72 dBm RF sensitivity was reported for a prototype wake-up receiver [47]. However, the modest sensitivity and OOK modulation scheme used by the wake-up radio restricts its span and reliability in a wireless link. Also, an off-chip bulk acoustic wave (BAW) RF preselect filter is required to reject potential interferers in the wake-up radio design, which increases the size and cost of the receiver. The super-regenerative receiver in [48] achieves high energy efficiency and sensitivity by using BFSK modulation. However, the narrowband receiver is susceptible to multipath fading or interference effects because the front-end operates at a fixed input frequency and is not tunable. The receiver relies on a narrowband RF pre-filter, and achieves a SIR of −10 dB at 10 MHz offset, but it degrades to 0 dB in-band. The UWB receiver in [49] also realizes good energy efficiency by duty cycling, but it requires synchronization to the received data, and possibly a wake-up receiver to detect incoming data. Synchronization in IR-UWB requires complex hardware and

Table 1.2 Receiver performance summary

Ref.	Year	CMOS Technology (nm)	Modulation	Sensitivity (dBm)	SIR at 10 MHz Offset	DC Power (mW)	Data Rate (Mbit/s)	Energy per Bit (nJ/bit)
[43]	2005	180	OOK	−100.5	N/A	0.4	0.005	80
[47]	2009	90	OOK	−72	−10	0.052	0.1	0.52
[48]	2010	180	BFSK	−86	−10	0.215	0.25	0.84
[49]	2007	90	UWB	−99	−15	35.8	0.1	2.5
[51]	2010	130	UWB	−55	−15	3.3	1.3	3.3

additional power consumption [50]. The other UWB receiver in [51] operates at 1.3 Mbit/s, but has poor sensitivity (−55 dBm) and is confined to RF inputs below approximately 1 GHz. The measured in-band SIR of the UWB receiver is higher at −15 dB, due to its spectral diversity. To be competitive, a non-duty-cycled FM-UWB receiver should consume less than 1 mW, has SIR better than −15 dB, has sensitivity better than −80 dBm, and has an energy efficiency of few nJ/bit.

A summary of existing FM-UWB transceivers reported in recent publications are listed in Tables 1.3 and 1.4 for transmitters and receivers, respectively. Unfortunately, it is clear that the energy efficiency is nowhere near the 10 nJ/bit (max.) required for an autonomous wireless system. The research work described in this book attempts to develop and demonstrate FM-UWB for low-power, low-data-rate applications such as WSN, WMBAN, and WPAN.

Table 1.3 FM-UWB transmitter performance comparison

Parameters	[52]	[53]	[54]	[55]	[56]
Year of publication	2006	2008	2009	2009	2011
Technology	180 nm CMOS	180 nm CMOS	180 nm CMOS	130 nm CMOS	180 nm CMOS
RF Tuning Range (GHz)	2.7–4.1	0.5–5	3–5.6	6.2–8.2	3.2–4.45
V_{DD} (V)	1.8	1.8	1.8	1.1	1.6
Phase Noise (dBc/Hz at 1 MHz)	−70	−75	−80.6	−107	−92
Max. Output Power (dBm)	−34	−9	−11	−5	−12.8
Bandwidth (MHz)	–	550	–	550	700
Sub-carrier Frequency (MHz)	–	1	–	1	51
Data rate (kbit/s)	100	100	–	100	1000
Power Consumption (mW)	7.2–14	2.5–10	19.8	4.6	18.2
Active Area (mm^2)	0.7	0.25	0.77	0.062	1.4
Energy Efficiency (nJ/bit)	72–140	70–100	198	46	18.2

Table 1.4 FM-UWB receiver performance comparison

Parameters	[57]	[58]	[59]	[60]
Year of publication	2006	2006	2009	2011
Technology	180 nm CMOS	SiGe BiCMOS 180 nm	SiGe BiCMOS 250 nm	180 nm CMOS
RF band, in GHz	3–5	3.1–4.9	7.2–7.7	3.4–4.3
Power Consumption, in mW	19.8	10	9.1	9.6
Receiver sensitivity, in dBm	−65	−46	−86.8	−70
Data rate, in kbit/s	100	100	50	50
Energy efficiency, in nJ/bit	198	100	182	192
Active area, in mm^2	–	0.72	0.88	1

The main challenge in power management is to maintain energy efficiency across different operating conditions and voltage levels. It also has to provide the wireless system with a clean supply voltage that has minimal ripple. A digital controller sets the power manager configuration based on conditions such as the availability of energy to be harvested, amount of required current load, requirement on step up/down voltage conversion, etc. The DC-to-DC converter in the power management sub-system is constrained by the chip area required for passive components and transistors that can be produced economically. Realizing on-chip passive devices, e.g., a nF-size capacitor or an inductor with a size of few μH [61], requires mm^2 of die area.

Several challenges facing the designer of on-chip autonomous wireless systems have been described. The work described in this book attempts to address many of the key barriers to implementation. Hardware solutions are proposed and demonstrated. The main objective is to realize a fully-integrated FM-UWB transceiver consuming less than 1 mW at an energy efficiency of better than 10 nJ/bit (see arrow trajectory in Figure 1.5). A power management sub-system will also be prototyped so that a battery and power scavenging devices can be used as energy sources.

1.3 Outline of this Book

This book is divided into six chapters. Chapter 2 describes the technical background relevant to wireless transceivers. Several wireless modulation schemes relevant to low-power wireless systems are described and compared, and the FM-UWB scheme is highlighted and analyzed in detail. A wireless link margin analysis and parameter-performance trade-offs for the transceiver are described in Chapter 2, along with specifications for the proposed autonomous transceiver. Conventional FM-UWB transceiver designs from the recent literature are also described and analyzed. The efficiency limitations of a wireless system are analyzed, and several CMOS circuit techniques from the literature that attempt to improve power efficiency are described.

A new architecture for a low-power FM-UWB wireless transmitter is proposed in Chapter 3. The fully-integrated transmitter prototype was implemented using 90-nm CMOS and tested as a proof of the concept. A current-controlled ring oscillator generates the RF carrier frequency, followed by a class-AB power amplifier (PA) to drive the antenna. A frequency calibration scheme utilizing the successive approximation technique and a frequency-locked loop (FLL) is also proposed and implemented. The transmitter has been characterized completely and the results validate the proposed architecture.

A wireless receiver design for FM-UWB using a regenerative RF preamplifier and an envelope detector is presented in Chapter 4. The receiver front-end was implemented in 65-nm CMOS, while baseband circuitry is implemented externally using off-chip components. The regenerative receiver is analyzed and compared (briefly) with a conventional FM-UWB receiver. Measured data for the new design are also compared to other low-power receivers.

A fully-integrated FM-UWB transceiver prototype based on an architecture that improves the energy efficiency, integration level, and manufacturability of an FM-UWB transceiver is the subject of Chapter 5. An RF-ICO generates a signal at one-third the operating frequency, which is then frequency multiplied by a tripler and amplified. The regenerative receiver front-end amplifier is implemented using inductive positive feedback followed by an envelope detector and IF amplifier. The transceiver was implemented in 90-nm CMOS and includes on-chip biasing, baseband modulator/demodulator, a serial input control port, and digital calibration circuitry.

Chapter 6 describes a power management circuit suitable for an autonomous FM-UWB system. Building blocks including DC-DC converters, battery, supercapacitor, and solar cell are described in the first sections of this chapter. The power management sub-system uses a button cell battery and a solar cell as energy sources. A supercapacitor stores energy during transient conditions. A hybrid of a switched-capacitor and an LDO regulator is designed to provide a constant supply voltage for the transceiver with an average conversion efficiency of 64%, and ripple of less than 0.1 mV in the output voltage.

References

[1] J. M. Rabaey, J. Ammer, T. Karalar, S. Li, B. Otis, M. Sheets, T. Tuan, "PicoRadios for wireless sensor networks: the next challenge in ultra-low-power design," in *IEEE ISSCC Dig. Tech. Papers*, Feb. 2002, pp. 200–201.

[2] A. Hamilton, "Time magazine's best invention of 2008: the internet of things." Available: http://www.time.com/time/specials/packages/0,28757,1852747,00.html.

[3] M. Chui, M. Loffler, R. Roberts, "The internet of things". Available on: http://www. mckinseyquarterly.com/The_Internet_of_Things_2538.

[4] D. Steel, "Smart dust," *UH ISRC technology briefing*. Available: http://www.bauer.uh.edu/uhisrc/FTB/Smart%20Dust/Smart%20 Dust.pdf

[5] W. Rhee, N. Xu, B. Zhou, Z. Wang, "Low power, non-invasive UWB systems for WBAN and biomedical applications," *International conference on ICTC*, Nov. 2010, pp. 35–40.
[6] S. Hussain, S. Schaffner, D. Mosychuck, "Applications of wireless sensor network and RFID in a smart home environment," *Proc. of communications networks and services research conference*, May 2009, pp. 153–157.
[7] "What is a smart building?," Available: http://www.smart-buildings.com.
[8] K. A. Townsend, J. W. Haslett, T. K. Tsang, M. N. El-Gamal, K. Iniewski, "Recent advances and future trends in low power wireless systems for medical applications," *Proc. System-on-Chip Real-Time Applications*, July 2005, pp. 476–481.
[9] B. Gupta, E. Cianca, M. Ruggieri, R. Prasad, "End to end vital sign monitoring system with FM-UWB technology," *International conference on devices and communications*, Dec. 2011.
[10] E. Cianca and B. Gupta, "FM-UWB for communications and radar in medical application," *Wireless Personal Communications, Springer,* Vol. 51, pp. 793–809, Dec. 2009.
[11] P. Pawelczak, R. V. Prasad, L. Xia, I. G. M. M. Niemegeers, "Cognitive radio emergency networks-requirement and design," *IEEE Symp. on new frontiers in dynamic spectrum access networks*, Nov. 2005, pp. 601–606.
[12] J. M. Rabaey, F. Burghardt, Y. H. Chee, et. al., "Short distance wireless, dense networks and its opportunities," *10th Euromicro Conference on Digital System Design Architectures, Methods and Tools*, Oct. 2007.
[13] ABI Research, "802.15.4 Wireless Sensor Networks: Market Data for 802.15.4." Available: https://www.abiresearch.com/market-research/product/1020117-802154-wireless-sensor-networks/
[14] Onworld, "Smart technology research, we provide world class business intelligence on emerging wireless markets." Available: http://www.onworld.com
[15] M. Rahy, "ULP meets energy harvesting: A game-changing combination for design engineers," *Texas Instrument white paper*, 2008.
[16] B. Jagannathan, R. Groves, D. Goren, et al., "RF CMOS for microwave and mm-wave applications," *Proc. of silicon monolithic integrated circuits in RF systems*, Jan. 2006, pp. 259–264.
[17] Z. Luo, A. Steegen, M. Eller, et al., "High performance and low power transistors integrated in 65 nm bulk CMOS technology," *IEDM Technical Digest*, pp. 661–664, Dec. 2004.

[18] P. H. Woerlee, M. J. Knitel, R. van Langevelde, et al., "RF-CMOS performance trends," *IEEE transaction on electron devices*, Vol. 48, No. 8, pp. 1776–1782, Aug. 2001.
[19] Revision of Part 15 of the Commission's Rules Regarding Ultra-Wideband Transmission Systems, First Report and Order, FCC 02-48, Federal Communications Commissions (FCC), 2002.
[20] ECC decision of 24 March 2006 on the harmonized conditions for devices using ultra-wideband (UWB) technology in bands below 10.6 GHz, C.E. document, 2006.
[21] "Technical specification ultra-wideband (UWB) devices." Available on: http://www.ida.gov.sg/doc/Policies%20and%20Regulation/Policies_and_Regulation_Level2 /IDATSUWB.pdf
[22] CEPT report to European Commission in response to: Harmonized radio spectrum use for ultra-wideband systems in the European Union, March 2005.
[23] J. F. M. Gerrits, M. Kouwenhoven, P. van der Meer, J. R. Farserotu, and J. R. Long, "Principles and limitations of ultra-wideband FM communication systems," *EURASIP Journal on applied signal processing*, pp. 382–396, 2005.
[24] J. F. M. Gerrits, H. Bonakdar, M. Detratti, et al., "A 7.2 −7.7 GHz FM-UWB transceiver prototype," *IEEE International conference on ultra-wideband*, Sept. 2009, pp. 580–585.
[25] J. F. M. Gerrits, J. R. Farserotu, J. R. Long, "Robustness and interference mitigation for FM-UWB BAN radio," *International symposium on medical information & communication technology*, March 2011, pp. 98–102.
[26] M. Hernandez, R. Kohno, "UWB systems for body area networks," *Proc. on ISSSTA*, Oct. 2010, pp. 112–115.
[27] IEEE Std 802.15.6-2012, IEEE standard for local and metropolitan area networks Part 15.6: Wireless Body Area Networks, Feb. 2012.
[28] R. Tandra, "Fundamental limits on detection in low SNR," *Master theses, UC Berkeley*, 2005.
[29] K. Vasanth, "Bipolar vs. CMOS: selecting the right IC for medical designs," *EE Times*, 2010. Available: http://www.eetimes.com/design/embedded/4211164/Bipolar-vs–CMOS–Selecting-the-right-IC-for-medical-designs.
[30] S. Malevsky, J. R. Long, "A comparison of CMOS and BiCMOS mm-wave receiver circuit for applications at 60GHz and beyond," *Analog Circuit Design*, Springer, pp. 327–342, 2008.

[31] R. Aigner, "SAW and BAW technologies for RF filter applications: a review of the relative strengths and weaknesses," *IEEE Ultrasonic symposium*, Nov. 2008, pp. 582–589.

[32] Li Shengyuan, S. Sengupta, H. Dinc, P. E. Allen, "CMOS high-linear wide-dynamic range RF on-chip filters using Q-enhanced LC filters," *IEEE International symp. on circuits and systems*, May 2005, pp. 5942–5945.

[33] P. Khademsameni, M. Syrzycki, "Manufacturability analysis of analog CMOS ICs through examination of multiple layout solutions," *Proc. symp. on defect and fault tolerance in VLSI systems*, 2011, pp. 3–11.

[34] J. Lansford, "Detect and avoid (DAA) for UWB: implementation issues and challenges," *International symp. on personal, indoor and mobile radio communications*, Sept. 2007, pp. 1–5.

[35] P. Galy, J. Jimenez, P. Meuris, W. Schoenmaker, O. Dupuis, "ESD RF protections in advanced CMOS technologies and its parasitic capacitance evaluation," *IEEE International conference on IC Design & Technology*, May 2011, pp. 1–4.

[36] CR2032 Lithium coin battery, http://www.microbattery.com.

[37] J. C. Jensen, R. Sadhwani, A. A. Kidwai, et al. "Single-chip WiFi b/g/n 1x2 SoC with fully integrated front-end & PMU in 90nm digital CMOS technology," in *Proc. IEEE radio frequency integrated circuits symposium*, May 2010, pp. 447–450.

[38] A. A. Emira, A. Valdes-Garcia, B. Xia, A. N. Mohieldin, A. Valero-Lopez, A. T. Moon, X. Chunyu, and E. Sanchez-Sinencio, "A dual-mode 802.11b/Bluetooth receiver in 0.25 μm BiCMOS," *in IEEE Int. Solid-State Circuits Conf. Dig. Tech. Papers*, 2004, pp. 270–271.

[39] W. Kluge, F. Poegel, H. Roller, M. Lange, T. Ferchland, L. Dathe, and D. Eggert, "A fully integrated 2.4 GHz IEEE 802.14.4-compliant transceiver for Zigbee^{TM} applications," *IEEE Journal of Solid-State Circuits*, Vol. 41, No. 12, pp. 2767–2775, Dec. 2006.

[40] J. Rousselot, J. D. Decotignie, "Wireless communication systems for continuous multi parameter health monitoring," *Proc. IEEE ICUWB*, Sept. 2009, pp. 480–484.

[41] J. R. Long, W. Wu, Y. Dong, Y. Zhao, M. A. T. Sanduleanu, J. F. M. Gerrits, and G. van Veenendaal, "Energy-efficient wireless front-end concepts for ultra-lower power radio," in *Proc. IEEE CICC*, Sept. 2008, pp. 587–590.

[42] D. C. Daly, A. P. Chandrakasan, "An energy efficient OOK transceiver for wireless sensor networks," *IEEE Journal of Solid State Circuits*, Vol. 42, No. 5, pp. 1003–1011, May 2007.

[43] B. Otis, Y. Chee, and J. Rabaey, "A 400 μW-RX 1.6 mW-TX super-regenerative transceiver for wireless sensor networks," in *IEEE ISSCC Dig. Tech. Papers*, Feb. 2005, pp. 396–397, 606.

[44] P. P. Mercier, D. C. Daly, and A. P. Chandrakasan, "A 19 pJ/pulse UWB transmitter with dual capacitively-coupled digital power amplifiers," in *Proc. IEEE radio frequency integrated circuits symposium*, June 2008, pp. 47–50.

[45] B. W. Cook, A. Berny, A. Molnar, S. Lanzisera, K. S. J. Pister, "Low-power 2.4-GHz transceiver with passive RX front-End and 400-mV Supply," *IEEE journal of solid state circuits*, Vol. 41, No. 12, pp. 2757–2766, Dec. 2006.

[46] J. Ayers, N. Panitantum, K. Mayaram, T. S. Fiez, "A 2.4 GHz wireless transceiver with 0.95nJ/b link energy for multi-hop battery free wireless sensor networks," *Symp. on VLSI circuits dig. tech. papers*, June 2010, pp. 29–30.

[47] N. M. Pletcher, S. Gambini and J. M. Rabaey, "A 52 μW, wake-up receiver with −72 dBm sensitivity using uncertain-IF architecture," *IEEE Journal of solid-state circuits*, Vol. 44, No. 1, pp. 269–280, Jan. 2009.

[48] J. Ayers, K. Mayaram, T. S. Fiez, "An Ultralow-Power receiver for wireless sensor networks," *IEEE Journal of Solid State Circuits*, Vol. 45, No. 9, pp. 1759–1769, Sept. 2010.

[49] F. S. Lee and A. Chandrakasan, "A 2.5 nJ/b 0.65 V 3-to-5 GHz subbanded UWB receiver in 90 nm CMOS," *IEEE Int. Solid-State Circuits Conf. Dig. Tech. Papers*, Feb. 2007, pp. 116–590.

[50] P. P. Mercier, M. Bhardwaj, D. C. Daly, and A. P. Chandrakasan, "A 0.55V 16Mb/s 1.6mW non-coherent IR-UWB digital baseband with ±1 ns synchronization accuracy," *IEEE Int. Solid-State Circuits Conf. Dig. Tech. Papers*, Feb. 2009, pp. 252–253.

[51] N. van Helleputte, M. Verhelst, W. Dehaene, G. Gielen, "A reconfigurable, 130 nm CMOS 108pJ/pulse, fully Integrated IR-UWB receiver for communication and precise ranging," *IEEE Journal of Solid State Circuits*, Vol. 45, No. 1, pp. 69–83, Jan. 2010.

[52] T. Tong, Z. Wenhua, J. Mikkelsen, T. Larsen, "A 0.18 μm CMOS low power ring VCO with 1 GHz tuning range for 3–5 GHz FM-UWB applications," *Proc. of* 10th *IEEE International Conference on Communication Systems*, 2006, pp. 1–5.

[53] A. Georgiadis, M. Detratti, "A linear, low power, wideband CMOS VCO for FM-UWB applications," *Microwave and Optical Technology Letters*, Vol. 50, No. 7, pp. 1955–1958, July 2008.
[54] A. Tsitouras, F. Plessas, "Ultra wideband, low-power, 3-5.6 GHz, CMOS voltage-controlled oscillator," *Microelectronic journal,* Vol. 40, pp. 897–904, 2009.
[55] M. Detratti, E. Perez, J. F. M. Gerrits, M. Lobeira, "A 4.6 mW 6.25–8.25 GHz RF transmitter IC for FM-UWB applications," *Proc. of the ICUWB*, Sept. 2009, pp. 180–184.
[56] B. Zhou, H. Lv, M. Wang, et al., "A 1 Mb/s 3.2-4.4 GHz reconfigurable FM-UWB transmitter in 0.18 μm CMOS," *Proc. IEEE RFIC Symposium*, June 2011, pp. 1–4.
[57] T. Tong, J. H. Mikkehen, T. Larsen, "A 0.18 μm CMOS implementation of a low power, fully differential RF front-end for FM-UWB based P-PAN receivers," *IEEE Singapore International conference on Communication systems*, Oct. 2006, pp. 1–5.
[58] J. F. M Gerrits, J. R. Farserotu, and J. R. Long, "A wideband FM demodulator for a low-complexity FM-UWB receiver," *Proceeding of the* 9th *European Conference on Wireless Technology*, Sept. 2006, pp. 99–102.
[59] Y. Zhao, Y. Dong, J. F. M. Gerrits, G. van Veenendaal, J. R. Long, and J. R. Farserotu, "A short range, low data rate, 7.2 GHz–7.7 GHz FM-UWB receiver front-End," *IEEE Journal of Solid State Circuits*, Vol. 44, No. 7, pp. 1872–1881, July 2009.
[60] B. Zhou, J. Qiao, R. He, et al., "A gated FM-UWB system with data-driven front-end power control," *IEEE Transactions on Circuits and Systems I*, Vol. 58, No. 12, Dec. 2011.
[61] C. Vaucourt, "Choosing inductors and capacitors for DC/DC converters," *Texas Instrument application report*, SLVA157, Feb. 2004.

2

Technical Background

2.1 Introduction

This chapter begins with a brief overview of well-known digital modulation technique for wireless communication. Subsequently, the conventional time-domain ultrawideband (UWB) and the frequency-modulated ultrawideband (FM-UWB) modulation scheme will be described in more detail. This Chapter will also describe a conventional FM-UWB transceiver and its technical specification. Finally, a list of CMOS circuit design techniques that enable low-power design will be described.

2.2 Digital Modulation

This section is focused on digital modulation techniques that are suitable for low-power wireless systems. Digital modulation of data for communication offers greater accuracy (e.g., fidelity, ability to add redundancy to correct error, etc.) than analog in the presence of noise and distortion. The main tradeoff among different types of modulation (e.g., ASK, PSK, FSK, etc.) is the required signal to noise ratio (SNR) for a certain receiver sensitivity, bandwidth efficiency, and complexity of the implementation.

Shannon's theorem states that channel capacity, C in bit/s, is an upper bound on the data rate that can be transmitted, as given by

$$C = B \cdot \log_2 \left(1 + \frac{S}{N}\right) = B \cdot \log_2 \left(1 + \frac{Eb}{N_O}\frac{R}{B}\right), \tag{2.1}$$

where a given average signal power, S, is transmitted through a communication channel that is subject to additive white Gaussian noise (AWGN) of power N [2]. The energy per bit is *Eb*, and N_O is the thermal noise spectral density. R/B, in bit/s/Hz is the ratio of data rate to signal bandwidth, also referred to as the spectral efficiency. As shown in Figure 2.1, modulation format such

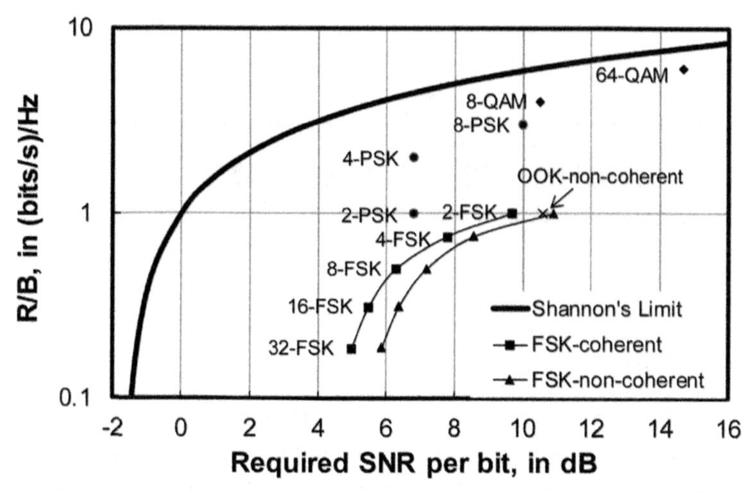

Figure 2.1 Spectral efficiency versus the required SNR per bit at BER = 10^{-3} for various modulation formats [3, 4].

as 64-QAM pack many bits of data into each signal transition, but requires higher SNR per bit for equivalent BER compared to a simpler modulation format. A complex modulation also requires extra hardware, such as a high-resolution ADC and DAC (for greater SNR), complex baseband circuitry (i.e., quadrature signal paths), and high-linearity PA that consumes more power. On the other hand, FSK modulation sacrifices bandwidth to achieve a lower SNR requirement. Although bandwidth can be traded to get better sensitivity (as in FSK modulation), both are bounded by the Shannon limit (see Figure 2.1). The asymptotic limit for SNR per bit as defined by Equation (2.1) is -1.6 dB, assuming that the bandwidth is infinite. An error correcting code (ECC) can reduce the required SNR by a few dB at the cost of reduced data throughput due to computational overhead of decoding the ECC which consumes more power [3].

For a low-power transceiver, the preferred modulation format should have low complexity and low SNR requirement, at the cost of bandwidth efficiency. Binary frequency shift keying (BFSK), binary phase shift keying (BPSK), and on-off keying (OOK) are typically the main choice of modulation format in low-power narrowband applications (see Tables 1.1 and 1.2).

OOK, whereby the carrier is turned on or off to represent digital bits is the simplest modulation format. For example, the transmitter can be turned off when a data bit is '0', making the transmitter power efficient. A digital coding scheme can be employed to maximize the "off" time of the transmitter [1].

The power amplifier for OOK can be implemented with a high-efficiency, non-linear amplifier. OOK modulated data may be received using an envelope detector, removing the need for a local oscillator and mixer. OOK typically uses a single frequency carrier, which simplifies the frequency synthesizer requirements. Two limitations of OOK modulation are 1) it is spectrally inefficient, and 2) it is susceptible to interferers. An OOK modulated carrier includes harmonic tones, which increases its effective bandwidth. Since the receiver detects the received bits by sensing energy within a specified band, any noise or interference can cause an error.

BFSK has a constant envelope signal. Hence, a high-efficiency, non-linear power amplifier can be used in the transmitter. A BFSK receiver typically uses a limiter, which reduces the susceptibility to fast fading. Non-coherent detection for BFSK is generally used at the cost of 1–2 dB less sensitivity compared to a (more complex) coherent demodulator [4].

2.3 Ultra-Wideband (UWB)

Wireless data transmission is defined as ultra-wideband (UWB) when the emitted signal bandwidth exceeds 500 MHz or 20% of its center frequency [5]. Unlike traditional wireless standards, where data is confined within a specific band, energy in a UWB signal is spread over a large portion of spectrum, giving it some immunity to multipath fading and interference. Transceivers operating in the unlicensed UWB range are constrained in their link span by FCC regulation in North America (see Chapter 1) to an average transmit power density of −41.3 dBm/MHz and peak transmit power of 0 dBm [6]. The RF power emitted by a UWB transmitter is consistent with the capabilities of an energy-efficient, short-range wireless transceiver. Co-existence with other UWB communication systems may be achieved by transmitting the signal within a selected portion of the bandwidth allocated for UWB systems, and employing detect-and-avoid or a frequency hopping scheme in a cognitive-type radio system [7]. A cognitive radio may change its parameters by monitoring channel conditions, (thereby avoiding interference), and thus allows efficient spectral reuse [8].

As seen from Equation (2.1), UWB offers greater capacity due to the large bandwidth available. It enables data throughput up to 1 Gbit/s across a short distance [9]. Employing UWB modulation for a low data rate application might seem counterintuitive. Fortunately, information spread over a large bandwidth benefits from an additional gain in the SNR called a processing gain [3, 10] (i.e., the ratio between bandwidth and data rate). As in other spread-spectrum modulation techniques (e.g., Code Division Multiple Access (CDMA)), the immunity to jamming for UWB at low data rate increases in proportion to the processing gain. Hence, specifications for the transceiver can be relaxed, e.g., lower transmitted output power or a higher receiver noise figure. This presents an opportunity to implement a simpler, lower power transceiver for low data rate applications.

UWB is not limited to a certain waveform, though a popular choice is a time-domain pulse [11]. The energy of the pulses that are emitted by the antenna occupies a large bandwidth. The impulse radio UWB (IR-UWB) transmitter originated early in the 20th century in the form of a spark-gap transmitter [12]. A typical transceiver (shown in Figure 2.2) is simple and power efficient. The transmitted data is modulated using pulse-position modulation (PPM), or BPSK. Figure 2.3 shows an example of an IR-UWB modulation scheme using BPSK. A burst of pulses is used in this example to

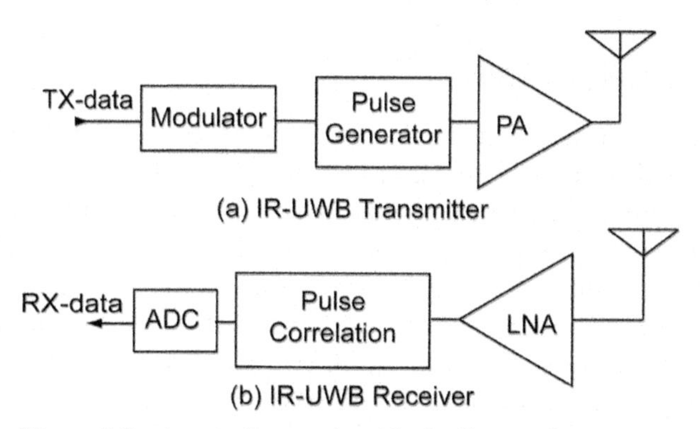

Figure 2.2 A typical transceiver blocks diagram for IR-UWB

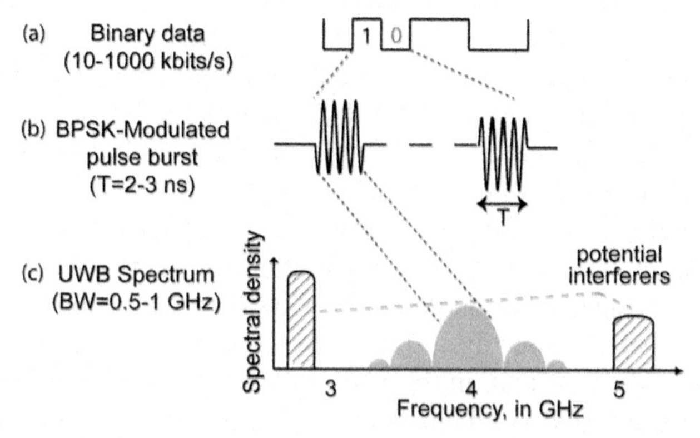

Figure 2.3 BPSK modulation scheme for low data rate IR-UWB [13].

boost the average output power without violating the maximum peak transmit power (as per FCC regulation [6]). Pulses produced by the generator are then transmitted by the antenna through a power amplifier. The received pulses are amplified and then demodulated using a correlation receiver. An ADC quantizes the demodulated signal and regenerates the received data bit stream.

There are several practical challenges that have prevented wide use adoption of IR-UWB transceivers. The transmitted pulses need to be accurately shaped such that the spectrum has a certain center frequency, bandwidth, and spectrum density. This is not an easy task because pulse shaping requires a complex RF filter that is sensitive to manufacturing variations. Typically, spectral side lobes emitted by an IR-UWB transmitter could interfere with other users. The power amplifier (PA), low-noise amplifier (LNA), and the antenna must have a wide bandwidth, and wideband amplifiers are typically less power efficient, although recent development has shown promising results for reducing LNA power consumption [14]. A wideband amplifier is also susceptible to in-band interference that could severely reduce the dynamic range. Another practical challenge for the receiver is blind timing synchronization at the receiver. An accurate timing reference and some time to acquire the correct synchronization are required.

Other types of modulation that utilize UWB spectra are emerging, such as orthogonal frequency-division multiplexing (OFDM), where data is transmitted using multiple narrowband carriers [15, 16]. OFDM transceivers need a complex carrier generator system that involves a fast PLL, and are neither simple nor low power. Thus, OFDM-type UWB transceivers are limited to high data rate applications. A novel UWB modulation scheme for low data rate applications that utilizes frequency modulation (FM) to produce a wideband transmit signal with a uniform spectral density is presented in the following section.

2.4 FM-UWB Modulation Scheme

Frequency-modulated ultra-wideband (FM-UWB) is well-suited to low-complexity, low-power wireless transceiver implementations, and transmission links up to 10 m [10]. FM-UWB uses a low modulation-index (m_i) FSK (e.g., $m_i = 1$) to encode binary data at a rate of 10–250 kbit/s (typical 100 kbit/s) onto a sub-carrier at an intermediate frequency (IF) of 0.5–2 MHz. This is followed by a high modulation index (e.g., $m_i = 500$) analog frequency modulation of the RF carrier, resulting in an RF signal bandwidth of 0.5–1 GHz, as shown in Figure 2.4. The RF carrier is swept back and forth at a

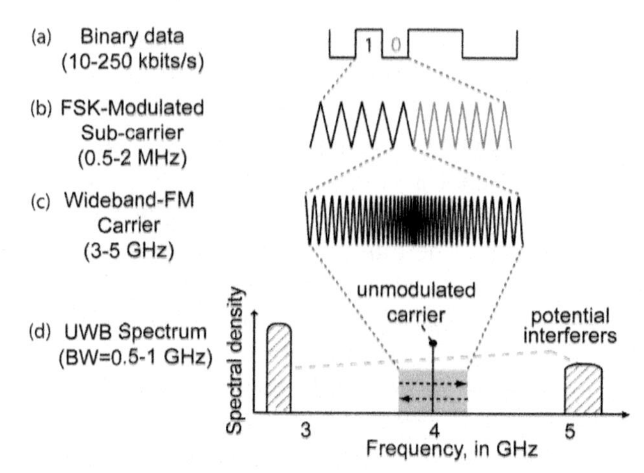

Figure 2.4 FM-UWB modulation scheme, where data (a) is FSK modulated onto a triangular sub-carrier (b). Subsequently wideband frequency modulation of RF carrier (c) results in UWB spectrum shown in (d).

constant rate. A uniform spectral density and steep spectral roll-off for the transmitted signal are realized through the use of a triangular sub-carrier waveform. The transmit bandwidth is determined by the sub-carrier amplitude, which can be controlled and calibrated easily. Not only the bandwidth, but also the center frequency is controllable, allowing flexibility in generation of the transmitted UWB spectrum.

The agility of FM-UWB also allows a more robust wireless system that avoids potential interferers by employing a frequency hopping scheme similar to frequency-hopped CDMA (FH-CDMA) [1]. However, a regular frequency hopping scheme is used instead of a random one. In this regard, FM-UWB has benefits associated with the spread-spectrum technique, such as immunity to jamming, a transmitting signal hidden in the background noise, and the ability to accommodate simultaneous transmissions in the same frequency band (i.e., up to 15 users) [17]. FM-UWB could also accommodate multiple users by using time-division multiple access (TDMA), frequency-division multiple access (FDMA), or sub-carrier FDMA sharing the same carrier frequency. In conclusion, FM-UWB offers robust, low-power, and low-cost access to unlicensed spectrum allocated for UWB that is available worldwide.

The FM-UWB transceiver operating frequency is defined by the regulators (see Figure 1.4), and preferably where maximum power could be transmitted in order to maximize link margin. The work described in this book will focus on the 3–5 GHz band. A modulated signal bandwidth of at least 500 MHz is required to conform to the definition of a UWB signal. Greater bandwidth means more signal spreading, which allows higher transmitted output power and link margin, but the transmitter consumes more power as a consequence. Flexibility to control the bandwidth is therefore advantageous for FM-UWB to maintain power efficiency when channel conditions vary.

Figure 2.5 shows various example of FM-UWB spectra for different sub-carrier IFs. The higher the IF, the more gradual the spectral roll-off and there will be peaking in the spectra. The sub-carrier IF needs to be at least one one-hundredth of the RF bandwidth to obtain a large modulation index, and create a uniform spectral density across the band. For the 3–5 GHz band, the IF should be limited to below 5 MHz, because above 5 MHz the spectrum is no longer uniform and power is concentrated at harmonics of the IF. Baseband data modulates the sub-carrier using BFSK modulation. Frequency deviation between symbols '0' and '1' needs to be at least equal to the data rate (i.e., modulation index $m_i > 1$) to avoid inter-symbol interference (ISI), which relaxes the FSK demodulator sensitivity requirement. A higher IF can be chosen to accommodate a higher data rate and a multi-user FDMA scheme [17]. The double modulation in the FM-UWB scheme limits the maximum data rate. A practical FSK demodulator, such as a PLL, requires a carrier at a higher frequency than the data rate (e.g., >10 times) [18]. FM-UWB could (theoretically) employ a maximum data rate of up to one-half of the minimum IF [19]. However, it needs an accurate FSK demodulator. The work described in this book focuses only on data rates less than 250 kbit/s, which are applicable to WSN, WPAN, and WMBAN applications.

FM-UWB uses direct modulation instead of up-converting the modulated signal as in a conventional radio. Therefore, the sub-carrier wave shape determines the spectral flatness and roll-off. Figure 2.6 shows the simulated FM-UWB spectra for various sub-carrier wave shapes. Triangular and saw-tooth sub-carriers have a uniform and flat spectrum density, respectively. Thus, they are suitable for FM-UWB. The triangular sub-carrier, which has a steeper roll-off and easier implementation is the preferred choice. Sine and hyperbolic sine functions create peaking at the edge of the signal band and steeper roll-off. As a consequence, the overall transmitted power must be reduced to conform to the spectral mask, which could reduce the overall link margin. A band-limited triangular or sawtooth sub-carrier also exhibits similar peaking, because carrier power is not distributed evenly across the bandwidth.

Non-linearity in the carrier generator (e.g., frequency versus voltage in a voltage-controlled oscillator) causes a non-uniform spectral density. Non-linearity only distorts the output spectral shape, and it is not critical for the

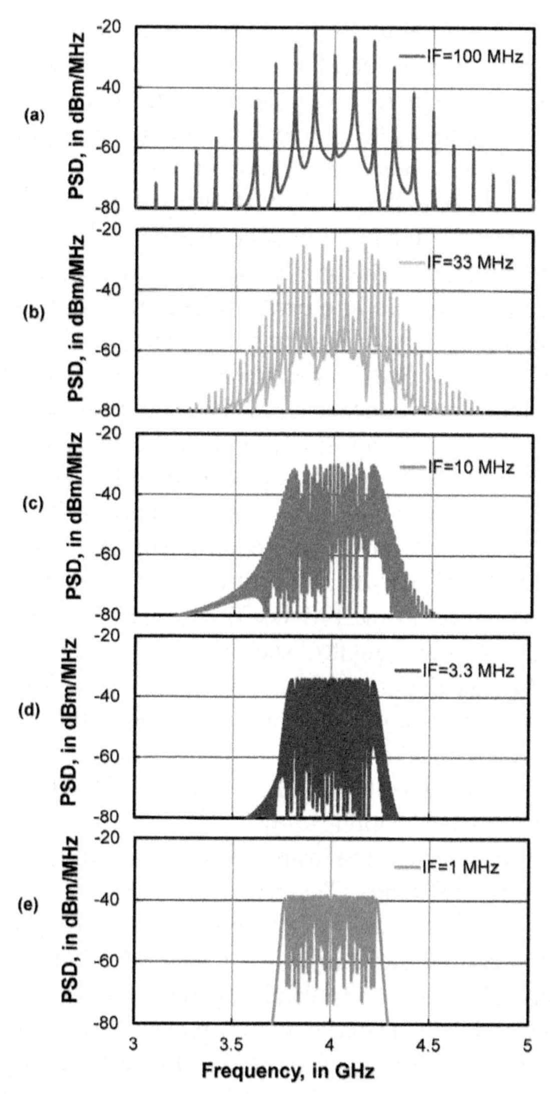

Figure 2.5 Simulated FM-UWB spectral density at 500 MHz bandwidth for different sub-carrier frequencies (IF).

FM-UWB communication. The output spectrum is usually distorted by the wireless channel condition (e.g., due to multipath), and is tolerable in an FM-UWB system [20]. Therefore, a linear carrier generator for FM-UWB is desired, but not necessary.

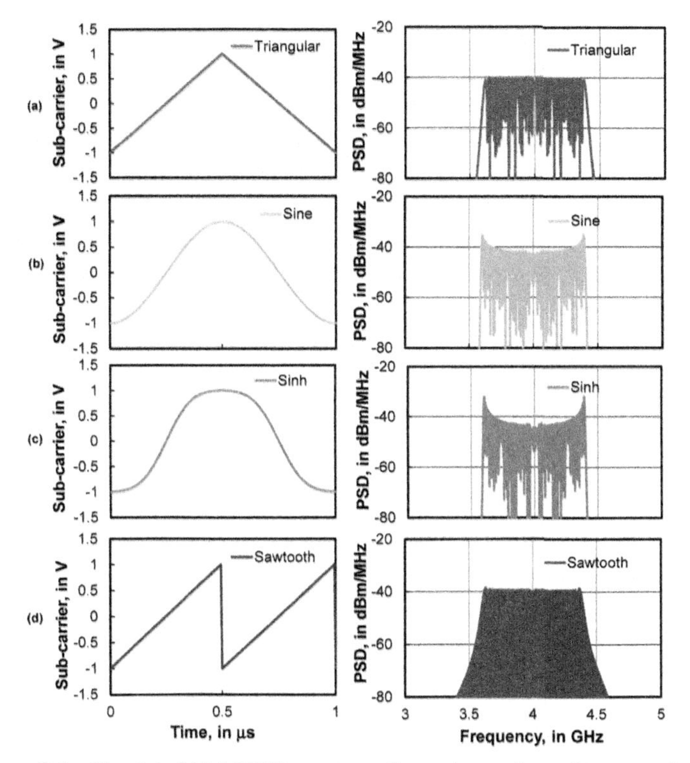

Figure 2.6 Simulated FM-UWB spectrum for various sub-carrier wave shapes.

Multipath occurs when the same transmitted signals arrive at the receiver through different paths and hence with different time delays. The phase difference between these signals adds constructively or destructively. This phenomenon typically occurs indoors, where walls reflect wireless signals and cause them to arrive at the receiver with different phases. Figure 2.7 shows a simulation setup for the multipath effect (2 paths) on an FM-UWB signal. Each path has a different time delay and magnitude. The attenuation

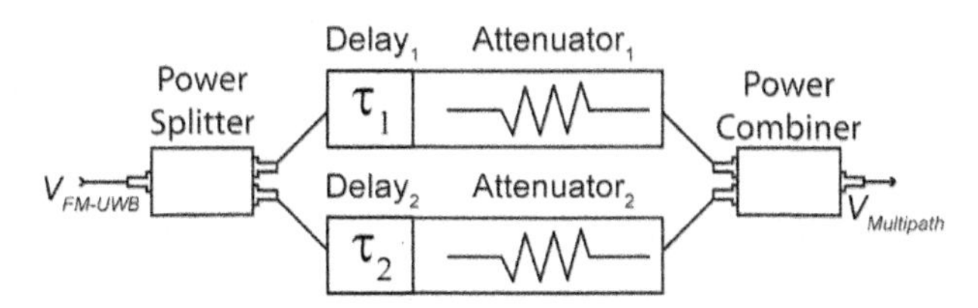

Figure 2.7 Setup used to simulate the multipath effect.

used in the simulation is based on the free space path loss (see Equation 2.4). Note that the signal is attenuated more for longer delays, emulating a longer path. Figure 2.8 shows that the simulated FM-UWB spectrum suffers from the two-path effect as the time delay differences vary. Multipath causes fading at specific frequencies that depend on the delay time difference. A shorter delay difference causes fewer in-band nulls, but they are wider and deeper in bandwidth. The fading could also be time varying, which makes it difficult to compensate. Fortunately, the transmitted power of an FM-UWB signal is spread across a wide bandwidth, hence fading does not significantly degrade the signal, unlike the narrowband FM radio where reception suffers when multipath fading occurs.

Alternatively, it is possible to use binary phase-shift keying (BPSK) to modulate the sub-carrier signal by using a sawtooth waveform to maintain

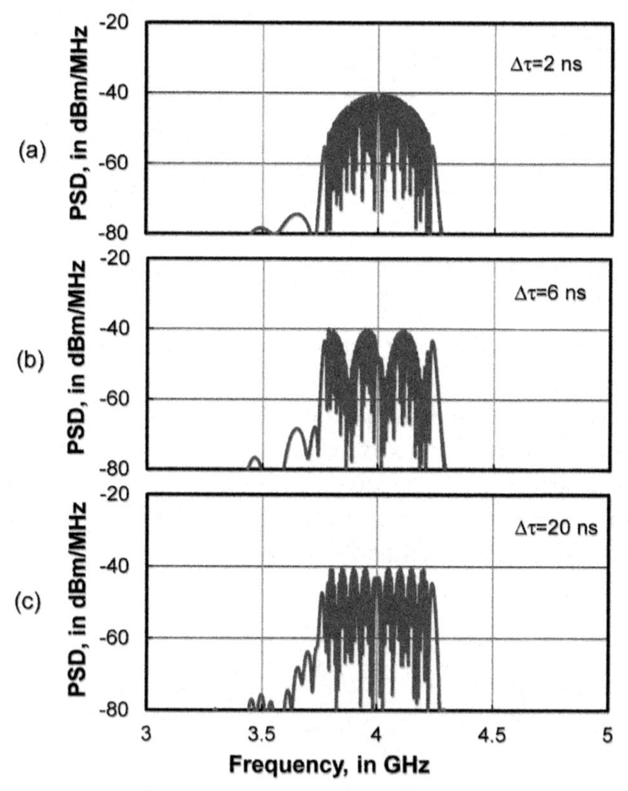

Figure 2.8 Simulated FM-UWB spectrum for two paths with a time delay difference of (a) 2 ns, (b) 6 ns, and (c) 20 ns.

a uniform output spectral density. The sawtooth showed in Figure 2.9(b) represents a BPSK modulated signal where two phases are represented by an increasing or decreasing ramp. The phase representing a '1' or '0' bit can be detected by a receiver employing a demodulator that distinguishes the direction of the ramp. BPSK modulation of the subcarrier opens up the possibility for higher data rates, or duty-cycled FM-UWB.

2.5 Specifications for FM-UWB Transceiver

A generic transceiver block diagram suitable for FM-UWB is shown in Figure 2.10. The FM-UWB receiver does not require a local oscillator or carrier synchronization. The transmitter (in its simplest form) may be implemented using just a wideband voltage-controlled oscillator, which results in a simple transceiver architecture that consumes little power in continuous operation. Digital data (e.g., from a sensor) modulates the sub-carrier oscillator. The

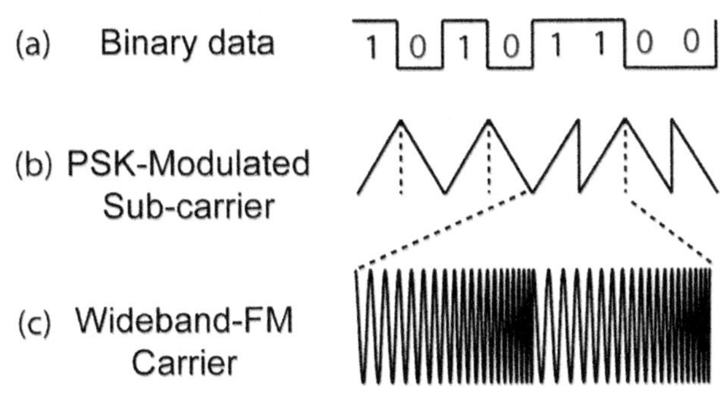

Figure 2.9 PSK sub-carrier signal using a sawtooth waveform.

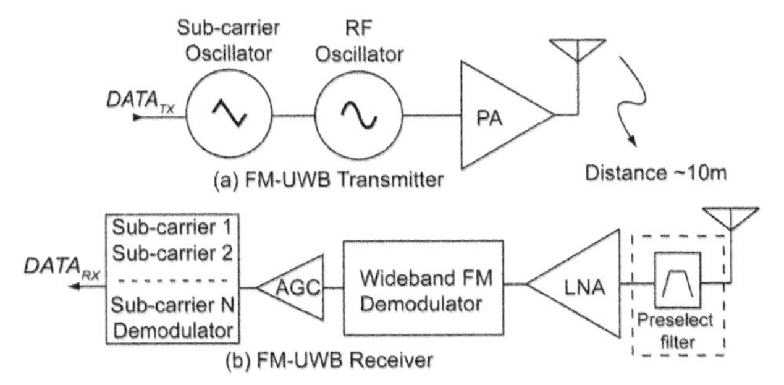

Figure 2.10 Block diagram of typical FM-UWB transceiver.

sub-carrier wave then modulates the RF carrier directly. A power amplifier (PA) drives the antenna. The receiver consists of an RF amplifier followed by a wideband FM demodulator. An automatic gain controlled (AGC) amplifier (or limiter) further amplifies the received sub-carrier. FSK demodulation of the received data intended for various users is then performed. Selectivity and robustness against out-of-band interference may be improved further by adding (optional) preselect filter between the antenna and RF input. The UWB antenna could also be modified to provide a frequency notch to mitigate expected interference [21]. Compared to a conventional radio [22], the transceiver for FM-UWB is much simpler and hence better suited to low-power applications.

The targeted specifications for the FM-UWB transceiver are summarized in Table 2.1. CMOS is the technology choice for the FM-UWB transceiver prototypes developed in this work. CMOS 65-nm and 90-nm technologies are employed [23], facilitated through MOSIS [24]. Standard transistors in 90-nm CMOS typically use a 1-V supply to avoid breakdown of the gate oxide. In the various prototypes described in subsequent chapters, a single 1-V supply powers the RF front-end, analog, and digital circuitry. A targeted power consumption of 1 mW is set for the receiver and transmitter to achieve an energy efficiency better than 10 nJ/bit at a data rate of 100 kbit/s. Chip area

Table 2.1 Target specification for the FM-UWB transceiver

Parameters	Values
Supply voltage, V_{DD}	1 V
Process technology	CMOS 65 or 90 nm
RF range	3–5 GHz
RF bandwidth	>500 MHz
Sub-carrier frequency	0.5–4 MHz
Data rate	<250 kbit/s
Power dissipation, at 4 GHz	<1 mW
Energy efficiency	<10 nJ/bit
Chip area	<1 mm^2
Transmitter	
Phase noise, at 4 GHz and offset frequency of 1 MHz	< -62 dBc/Hz
Output power	> -14.3 dBm
Spectral flatness within 500 MHz	<3 dB
Receiver	
Receiver sensitivity	< -80 dBm
Bit-error rate	10^{-3}
$\|S_{11}\|$	< -10 dB
Signal-to-interference ratio (NBI at 4 GHz)	< -20 dB

is minimized by using as few on-chip inductors as possible and the minimum number of bondpads. A chip area of 1 mm^2 is the goal for the full transceiver.

The RF carrier frequency deviation in an FM-UWB system is typically greater than 250 MHz, which implies that phase or frequency jitter of the transmit source has less effect upon the received signal-to-noise ratio (SNR) than for a narrowband system. It has been shown that FM-UWB is able to tolerate transmitter phase noise as high as −73 dBc/Hz at 1-MHz offset with no significant degradation in bit-error performance [25]. The FM-UWB phase noise requirement is much relaxed compare to UWB-OFDM, which requires −87 dBc/Hz at 1-MHz offset [26]. RF carrier deviation for the FM signal compared to FM noise has to be higher than the required SNR. For a non-coherent BFSK demodulator at a BER of 10^{-3}, the required SNR is 11 dB [10]. FM-UWB has a minimum frequency deviation of 250 MHz, which corresponds to maximum frequency jitter due to white noise, σ_f of 49 MHz. Obviously, the higher the modulated FM bandwidth, the higher the tolerable frequency jitter is. Frequency jitter can be related to time jitter, σ_T by the equation [27]

$$\sigma_T = \frac{\sigma_f}{f_C^2}, \tag{2.2}$$

where f_C is the carrier center frequency. Equivalent phase noise at frequency offset Δf can be described as [27]

$$L(\Delta f) = \frac{f_C^3}{\Delta f^2}\sigma_T^2. \tag{2.3}$$

From Equation (2.3) and the calculated frequency jitter requirement, the FM-UWB transmitter must have a phase noise better than −62 dBc/Hz at 1-MHz offset to satisfy the minimum SNR for the BFSK at BER of 10^{-3}.

The spectral mask for UWB signals dictates that the maximum effective isotropic radiated power (EIRP) density is −41.3 dBm/MHz. For a 500-MHz modulation bandwidth, the power density limit translates into a total carrier power of −14 dBm, or 40 µW. A margin of 4 dB is added to the PA output in order to compensate for losses in chip packaging, the antenna matching network and variations in the antenna gain. Spectral flatness indicates how uniform the in-band signal spectral density is, and it is limited to a maximum variation of 3 dB.

Assuming line of sight (LOS) transmission, path loss (PL) between transmitter and receiver at a short distance can be approximated by the Friis' free space equation

$$PL = \left(\frac{4\pi d}{\lambda}\right)^2, \tag{2.4}$$

where d is the span between the transmitter and receiver, and λ is wavelength. Equation (2.4) is accurate when the span is higher than one wavelength (7.5 cm for $f_C = 4\,\text{GHz}$). At shorter distances, the antenna operates in the near field, and an advanced path loss model must be used [28]. Friis' equation is adequate in practice for calculating distances between the transmitter and receiver. A typical indoor FM-UWB link that covers a distance of 10 m translates into a free space propagation loss of 66 dB at 5 GHz (assuming 0-dBi antenna gain). Thus, the sensitivity of the receiver required to accommodate a link span of 10 m is at least -80 dBm, as listed in Table 2.1.

Sensitivity of the receiver (P_{RX_MIN}) can be calculated using the link budget equation as follows

$$\text{P}_{RX\_MIN} = 10\log(kTB_{RF}) + NF + SNR_{FSK} - G_P. \tag{2.5}$$

The 50-Ω thermal noise in a 500-MHz bandwidth corresponds to -87 dBm noise power, which gives 7-dB SNR at the input of a receiver 10 m away from the transmitter. The required SNR at the receiver baseband for demodulation of 2-FSK at a bit-error rate (BER) of 10^{-3} is 11 dB. FM-UWB may be considered a spread spectrum system with a processing gain (G_P) determined by the ratio of the RF to data bandwidths. Assuming 100 kbit/s data rate, the bandwidth of the 2-FSK data signal is 200 kHz, giving a processing gain of 34 dB (ideally). However, the receiver output SNR is a quadratic function of the RF-input SNR due to the non-linearity inherent in this type of non-coherent receiver. This results in an additional 6 dB degradation in SNR [10], so the overall margin for the receiver noise figure is the receiver input SNR (6 dB), minus the sum of the receiver penalty of 6 dB and required SNR_{FSK} of 11 dB, plus the processing gain (34 dB), or 23 dB margin. The margin will be reduced by the non-ideality of the filter and losses from envelope detector (if it is used). At a higher data rate, the sensitivity reduces because the 2-FSK data bandwidth increases and thus G_P decreases.

RF input matching (i.e., $|\text{S}_{11}| < -10\,\text{dB}$) is required to minimize energy reflection between the antenna and receiver. The sensitivity to single-tone interference is defined by detecting the degradation in BER up to 10^{-3}. Signal-to-interference ratio (SIR) measures the strength of the input signal compared to interference, and is negative when the interference power is stronger than input signal. A UWB radio should be able to withstand SIR of less than -20 dB from narrowband interference (NBI) [29]. The following section will review the previously reported work on FM-UWB transceivers.

2.6 Conventional FM-UWB Transceiver

The existing FM-UWB transmitters and receivers are usually based on the RF front-end shown in Figure 2.11. The conventional FM-UWB transmitter of Figure 2.11(a) [30, 31] generates the RF carrier using an LC-oscillator. An LC oscillator is widely used in RF wireless transceivers because of its simplicity, low power consumption at high frequencies of oscillation, and low phase noise. The frequency of the oscillator depends on the resonant frequency of its LC (inductor and capacitor) tank,

$$f_C = \frac{1}{2\pi\sqrt{LC}}. \tag{2.6}$$

The losses of the tank must be compensated by a negative impedance generated by an active device. Therefore, the tank Q-factor, especially for an on-chip inductor with $Q < 20$, limits the minimum power consumption of the LC

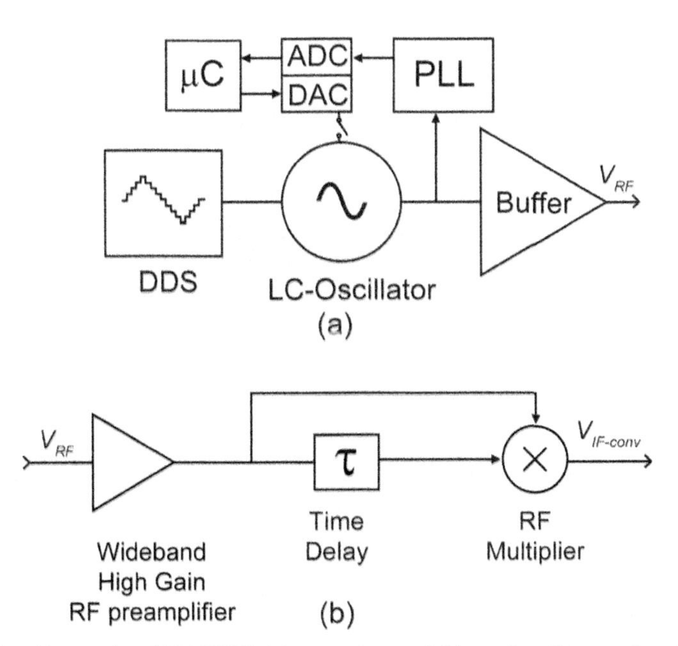

Figure 2.11 Conventional FM-UWB (a) transmitter and (b) receiver front-ends architectures.

oscillator [32]. An LC-oscillator that has a tuning range less than 20% of the center frequency might not have enough frequency range for the FM-UWB transmitter. The oscillator output drives a 50-Ω antenna load using a CMOS buffer. To calibrate its center frequency, a periodic duty-cycled PLL plus external ADC/DAC and a microcontroller are included. The loop is closed until the oscillator reaches the desired frequency, and then it is opened for the remaining time period of the calibration cycle.

The sub-carrier could be generated by a direct digital synthesizer (DDS) and DAC [33]. The clock frequency of the DDS must be at least 24 times the sub-carrier IF. The minimum resolution of the DAC is defined by the RF bandwidth of 500 MHz divided by a resolution bandwidth of 1 MHz, which is 500, or 9-bits. The simulated power consumption of the DDS and DAC is 0.75 mW [33].

The conventional FM-UWB receiver shown in Figure 2.11b [30] first amplifies the FM signal using a wideband LNA and then transforms FM to a phase-modulated (PM) signal via a delay line with a constant time delay (τ). The PM signal is then multiplied with the original FM signal to yield an amplitude-modulated (AM) output (illustrated in Figure 2.12a). By removing the high-frequency carrier using a lowpass filter, the transmitted sub-carrier signal ($V_{IF\text{-}CONV}$) is recovered.

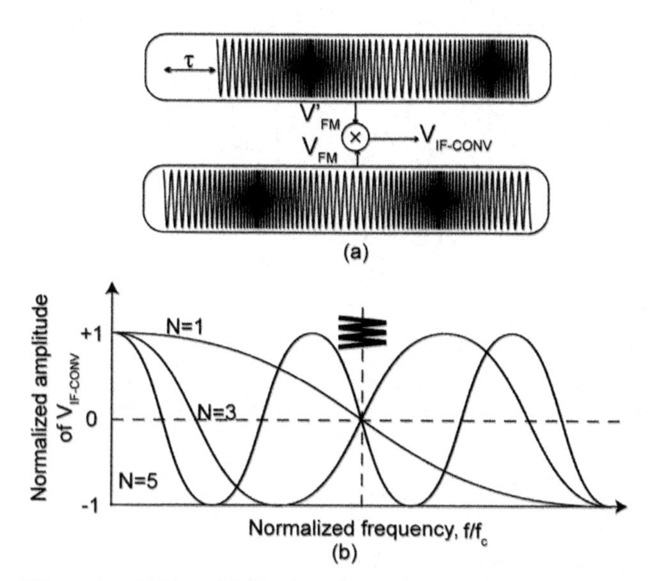

Figure 2.12 Illustration of (a) multiplication of an FM-UWB signal with its delayed version, and (b) its transfer function between amplitude and frequency.

Assuming an ideal multiplier with gain of K and an ideal lowpass filter, the demodulated voltage as a function of frequency can be expressed by

$$V_{IF-CONV}(f) = K\frac{V_{FM}.V'_{FM}}{2}\cos(2\pi f\tau). \tag{2.7}$$

The delay (τ) must be an odd multiple of quarter periods at the center frequency of carrier signal, as expressed by

$$\tau = N\frac{T}{4} = \frac{N}{4f_c}, \quad \text{for } N = 1, 3, 5, \ldots \tag{2.8}$$

By substituting Equation (2.8) into (2.7), $V_{IF\text{-}CONV}$ can be expressed as

$$V_{IF\text{-}CONV}(t) = K\frac{V_{FM}.V'_{FM}}{2}\cos\left(N\frac{\pi}{2}\frac{f_c \pm \Delta f_c(t)}{f_c}\right). \tag{2.9}$$

Equation (2.9) predicts that the bandwidth of the FM signal $(2\Delta f_c)$ determines the amplitude of the demodulated signal. As illustrated in Figure 2.12b, different values of N produce different transfer functions. N can be chosen to obtain maximum sensitivity and voltage swing based on the bandwidth of the signal. The required delay can then be derived from N and f_C. A wideband delay would consume a large inductor area [34] or large power consumption [35] if implemented on a chip. Additionally, some tuning elements are required to tune the delay to its optimum value when the center frequency or operating frequency band is changed.

Unfortunately, the conventional transceiver architecture implementation did not achieve good enough power efficiency for the proposed autonomous wireless system (see Tables 1.3 and 1.4). Therefore, a new architecture is developed in this work to realize a low-power FM-UWB transceiver suitable for autonomous applications. The following section will briefly describe various methodologies and circuit techniques that could improve power efficiency in CMOS technology.

2.7 Survey of Low-Power CMOS Circuits

Circuit design techniques or methodologies are often developed to overcome CMOS technology limitations. As CMOS technology advances into smaller feature sizes, it allows compact circuits and higher integration. However, there are several factors that degrade analog circuit performance as technology advances, such as, lower supply voltage, lower transistor breakdown voltage, and lower transistor intrinsic gain, *gm.ro*. There is also a higher mismatch between transistors and gate leakage current [36]. A survey of various state-of-the-art CMOS circuits suggests several design methodologies and techniques that can be adopted in the FM-UWB transceiver design to improve performance and reduce power consumption. Key guidelines to follow when designing low-power circuits are:

- Bias CMOS in the sub-threshold region [37].
- Increase amplifier gain by improving its output impedance (e.g., gain boosting and Q-enhancement techniques [38, 39]).
- Maximize headroom, i.e., use lowest possible V_{DD}, or apply the current-reuse technique [14, 40].
- Manage power by adjusting current consumption or shutting down circuits periodically (i.e., duty cycling) [41, 42].
- Design systems that require relaxed subsystem performance or fewer building blocks for a given functionality (e.g., low complexity and non-coherent radio) [10, 43].
- Use digital circuits and calibration to compensate for non-idealities (e.g., offset, mismatch, non-linearity, frequency error, or temperature drift) [44, 45].

Note that not all of these principles are applicable to the FM-UWB transceiver design. This list only offers guidelines that may be utilized when there is an opportunity.

2.7.1 Sub-Threshold CMOS

Transconductance (*gm*) is directly proportional to voltage gain and bandwidth in amplifier circuits. The transconductance per unit of current (i.e., transconductance efficiency, *gm/Id*) of a transistor is lower when it operates at a higher current density (see Figure 2.13) [37]. Small-signal, unity-current-gain frequency (or transit frequency, f_T) cannot be arbitrary small for the device to operate at RF. Sub-threshold bias is a powerful tool when reducing power consumption, but as can be seen from Figure 2.13, the device is slower (i.e., operates at lower f_T). On the other hand, the CMOS transistor can be biased at maximum f_T but at the cost of power efficiency (i.e., requiring more current for a certain *gm*). As CMOS is scaled further and f_T increases for shorter gate lengths in CMOS, sub-threshold-biased circuits should become a viable technique for RF-circuits implementation.

Figure 2.13 shows gm/I_D and f_T versus current density of an NMOS device (width of 1 μm and length of 90 nm) in CMOS 90-nm technology.

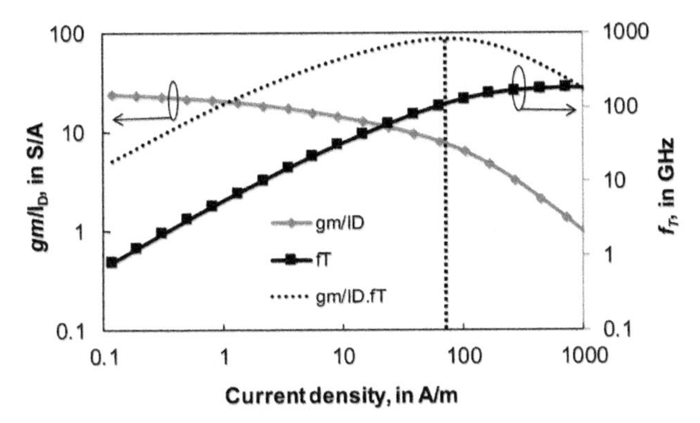

Figure 2.13 gm/I_D and f_T versus current density of NMOS device in 90 nm CMOS (W = 1 µm, L = 90 nm).

The current density at the peak f_T of 160 GHz for a 90-nm NMOS device is 0.8 mA/µm. The optimum trade-off (i.e., maximum product of gm/I_D and f_T) can be achieved at current density of around 80 µA/µm [46].

2.7.2 Gain Boosting or Q-Enhancement Technique

Gain boosting is a technique that increases the voltage gain of a single-stage amplifier by increasing its output impedance. This technique is suitable for low-power circuits because the cost of boosting is smaller than the power consumed by adding extra stages. At low frequency, output impedance can be boosted by adding an auxiliary amplifier in the negative feedback configuration [38]. Typically, a passive inductor is used at the output in a resonant tank to provide an output impedance in the order of a few kΩ at RF. Positive feedback could also be used to boost the output impedance in the resonant tank (i.e., Q-enhancement), at the risk of instability [39].

A simplified RF common-source amplifier (shown in Figure 2.14) has a voltage gain that can be described as

$$A_V = gm_n \times Z_L, \tag{2.10}$$

where gm_n is the transconductance of the NMOS transistor ($\mathrm{M_N}$), and Z_L is the impedance of the LC tank. The amplifier has a bandpass response, where maximum gain is attained at the resonant frequency of the tank. At RF, the on-chip inductor usually determines the tank Q-factor. Many efforts have been made to improve the quality factor for an inductor on-chip [47]. On-chip inductors of several nH and Q of more than 10 in the GHz range are

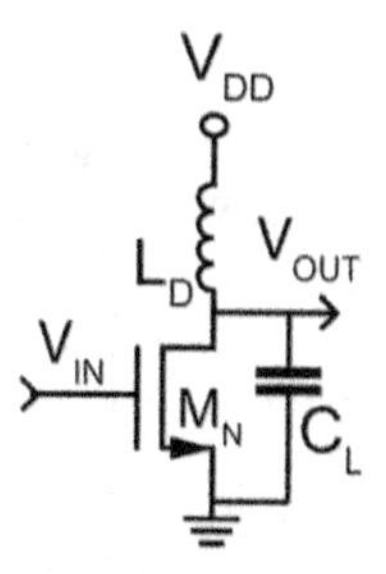

Figure 2.14 Schematic of a single stage RF amplifier.

now practical. Losses can be approximated as a lumped resistor. A negative resistor synthesized using an active device will cancel this loss. Effectively, the Q of the lossy component is enhanced by the active circuit at the cost of power consumption. The negative resistance is realized using a feedback that reverses the signal phase (i.e., positive feedback) [49].

A radio-frequency receiver typically receives a weak incoming signal, hence the need for amplification at RF. The Q-enhancement technique boosts the gain at low bias currents by operating the amplifier on the verge of instability. For example, the closed-loop transfer function of the positive-feedback amplifier shown in Figure 2.15(a), with forward path gain A and feedback factor β, is given by

$$H_{cl} = \frac{A}{1 - A\beta}. \tag{2.11}$$

The gain increases as loop gain ($A\beta$) approaches unity, but the bandwidth decreases as the gain-bandwidth product is approximately constant, as illustrated in Figure 2.15(b) [50].

Such a regenerative amplifier (i.e., an amplifier with positive feedback) is employed in Armstrong's regenerative FM radio receiver shown in Figure 2.16. Here, the FM signal is received by the antenna, and transformer

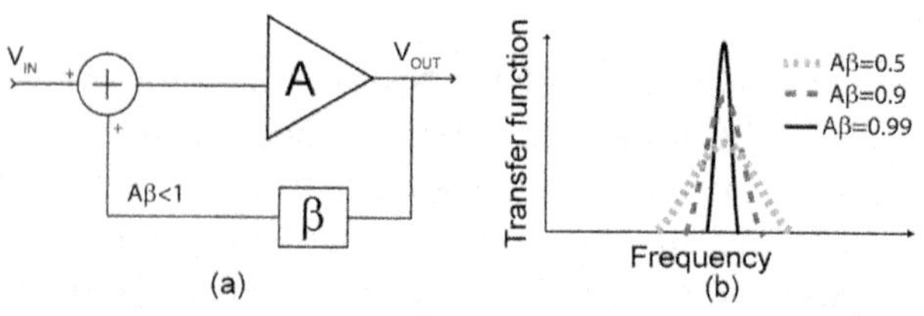

Figure 2.15 Positive feedback (a) block diagram and (b) transfer function.

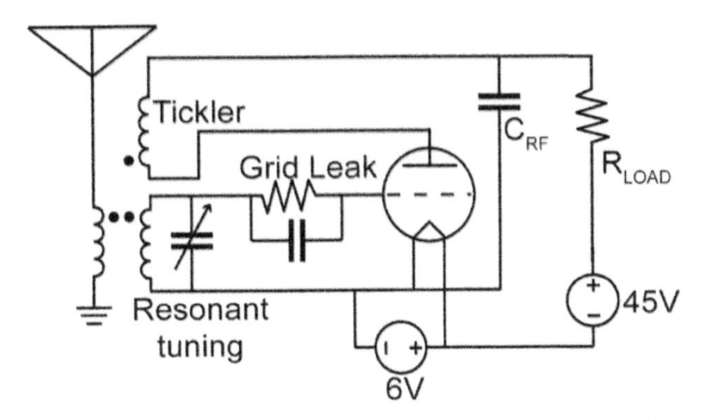

Figure 2.16 Armstrong's wireless FM receiving system [48].

coupled to the grid (analogous to the CMOS transistor gate) of a tube. The output signal is fed back through a 'tickler', which is coupled to the input transformer in a positive feedback fashion. By adjusting the physical distance between the tickler and coupling transformer coils, the positive feedback and hence the gain of the amplifier can be varied.

2.7.3 Maximum Voltage Supply Utilization

In low-power circuits, the available voltage or "headroom" is a precious resource. A single supply voltage is usually used, and all the circuits on the IC must utilize it. Individual circuits block may have different optimum supply voltages, but the highest voltage requirement usually sets the overall supply. This compromises efficiency as not all the circuits utilize the full headroom available. However, a multiple voltages supply would increase complexity, size, and cost of the power management sub-system required for implementation.

A series connection (e.g., cascoding) of circuit blocks is an obvious way to reduce current consumption, as illustrated in Figure 2.17. However, the designer must ensure that unwanted coupling between blocks is avoided and also regulate the voltage supplying each sub-circuit connected between the supply and ground. For example, a 2-stage RF amplifier can re-use the bias current consumed in both amplifier stages by cascoding the bias path [14]. The signal and DC paths are separated by inductor/capacitor networks. In [40] the VCO uses the bias current of other RF blocks to save overall power consumption. Stacking more than one circuits is possible (a VCO and a 2-stage PA [40]), but requires circuit topologies that operate properly at a low supply voltage. Ultra-low voltage ($<$0.5 V) circuit techniques are useful

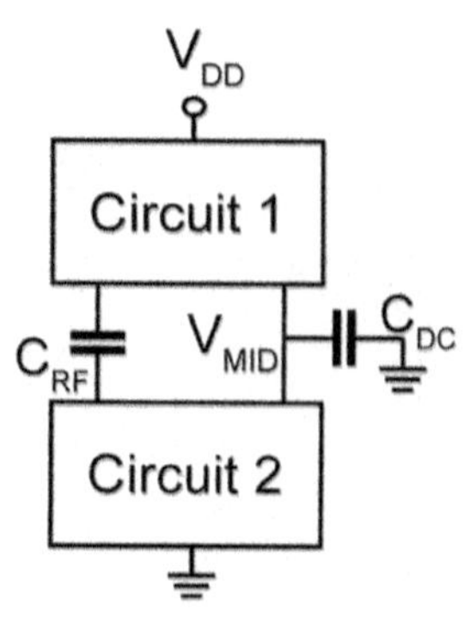

Figure 2.17 Illustration of current reuse for 2 circuit blocks.

if they can be stacked easily to ultimately reduce power consumption. A CMOS RF receiver, analog filter and opamp topology for ultra-low supply voltages have been proposed in [51, 52].

2.7.4 Adaptive Power Control

Adaptive power control is an obvious technique that leads to reduced power consumption for the transceiver circuit. There are various ways of controlling the power that can be classified as duty cycling and adaptive biasing. Duty-cycling is turning on and off the transceiver periodically to scale power consumption with respect to data rate, available data, or power source. Adaptive biasing changes the bias level in the transceiver according to channel conditions, e.g., reducing the receiver sensitivity and/or transmitted power when the received signal is strong.

A higher data rate transceiver usually has the advantage in term of energy efficiency per bit [53]. Duty cycling assumes that the data rate could be scaled easily without any penalty (i.e., to maintain energy efficiency). However, there are challenges when applying duty cycling in the transceiver, as extra time is required between on and off states to prepare the supply, bias, reset, etc. Hence, duty cycling at rates less than the order of once per microsecond is usually not effective. Reliability of transmission for a concentrated data burst is a problem if fading occurs. This is aggravated at very low duty cycles ($<0.1\%$), where all the data transmitted in a very short time. There is also a delay (latency) when duty cycling is applied that could pose a problem in medical applications, for example.

There is also a need for time synchronization or rendezvous between a duty-cycled transmitter and receiver (i.e. transmitting and receiving in agreed time). Several rendezvous schemes have been proposed in [54] that complicate the system as they require a clock accuracy better than 1% [55] and might require an extra wake-up receiver. The rendezvous scheme is also prone to channel fading [54].

Adaptive biasing schemes that change the dynamic range based on the channel conditions could be applied in the transceiver to reduce power consumption [56]. Also, circuits like the class-AB amplifier that change the bias current automatically with the signal amplitude could be used [57]. There is a trade-off between DC power consumption and sensitivity in the receiver and output power in the transmitter. Hence, optimum power consumption is realized at a certain link span [58].

2.7.5 Low Complexity Transceiver Architecture

Low power can also be achieved from simpler hardware at the cost of a performance, such as lower data rate or sensitivity. For example, a receiver that does not need a low NF could omit the LNA [59]. A super-regenerative scheme simplifies the receiver design by only using one oscillator [60]. Alternatively, a modulation scheme might be devised that accommodates simpler hardware such as the aforementioned FM-UWB modulation scheme [10].

Hardware requirements could also be simplified using non-coherent detection. Coherent detection implies that the receiver must know the carrier phase, and this information is obtained through synchronization. Tracking the received RF signal phase requires additional hardware (i.e., using a PLL), and implies higher power consumption [61]. Non-coherent detection is simpler and can consume less power because it does not require a local oscillator.

In some modulation schemes, the carrier phase is irrelevant and therefore phase locking to the carrier is not necessary. In this case, non-coherent detection is possible although there is some penalty in sensitivity. It is generally thought that performance of a coherent receiver is superior to non-coherent in a typical additive, Gaussian white noise environment. In reality, wireless communication signals are affected by fluctuations in amplitude and phase due to frequency-selective fading. Additionally, sensitivity is not critical in short links such as a WPAN.

Some advantages of non-coherent detection are [62]:

1. Better performance in fast frequency selective fading, large Doppler spread, strong phase noise, or co-channel interferences scenarios.
2. No time delay for resynchronization
3. Simpler hardware (no PLL)

There are 3 categories of non-coherent detection [62]:

1. Energy detector (e.g., OOK system using envelope detector)
2. Orthogonal signaling (e.g., FSK receiver, where signals are orthogonal irrespective of carrier phase)
3. Differential signaling (use the current carrier phase to decode the phase difference relative to the next symbol).

2.8 Summary

In this chapter, wireless modulation schemes suitable for low power consumption were described. Arguments outlining the advantage for an autonomous wireless device to utilize the available, unlicensed UWB spectrum were presented. The FM-UWB scheme with its advantages and limitations will therefore be the main focus of this book. The emphasis is on realizing an energy-efficient FM-UWB transceiver that satisfies the specifications listed in Table 2.1, that improved upon existing FM-UWB transceivers.

Various low-power circuit techniques for CMOS technology have been summarized and will be utilized where possible. For example, the possibility of using a regenerative amplifier along with an envelope detector in the receiver will be investigated. This results in a narrowband receiver that is simpler, but requires wideband tuning to process a UWB signal. The following chapter will describe an FM-UWB transmitter prototype realized in 90-nm CMOS.

References

[1] Q. Tang, S. K. S. Gupta, L. Schwiebert, "BER performance analysis of an on-off keying based minimum energy coding for energy constrained wireless sensor application," *IEEE International Conference on Communication*, Vol. 4, 2005, pp. 2734–2738.

[2] C. E. Shannon, "A mathematical theory of communication," *The Bell System Technical Journal*, Vol. 27, pp. 623–656, Oct. 1948.

[3] J. G. Proakis, *Digital communications*, McGraw Hill Inc., 1995.

[4] B. P. Lathi, *Modern digital and analog communication systems*, Oxford University Press, third ed., 1998.

[5] L. Yang, G. B. Giannaki, "Ultra wideband communication - an idea whose time has come," *IEEE Signal Processing Magazine*, pp. 26–54, Nov. 2004.

[6] Revision of Part 15 of the Commission's Rules Regarding Ultra-Wideband Transmission Systems, First Report and Order, FCC 02-48, Federal Communications Commissions (FCC), 2002.

[7] J. Lansford, "Detect and avoid (DAA) for UWB: implementation issues and challenges," *International symposium on Personal, Indoor and Mobile Radio Communications*, Sept., 2007, pp. 1–5.

[8] A. Sahai, R. Tandra, S. M. Mishra, and N. Hoven, "Fundamental design tradeoffs in cognitive radio systems," *proceeding of TAPAS*, Vol. 222, No. 2. Aug. 2006.

[9] S. Iida, K. Tanaka, H. Suzuki, N. Yoshikawa, N. Shoji, B. Griffiths, D. Mellor, F. Hayden, I. Butler, and J. Chatwin, "A 3.1 to 5 GHz CMOS DSSS UWB transceiver for WPANs," in *IEEE Dig. Tech. Papers*, Feb. 2005, pp. 214–215.

[10] J. F. M. Gerrits, M. H. L. Kouwenhoven, P. R. van der Meer, J. R. Farserotu, and J. R. Long, "Principles and limitations of UWBFM communications systems," *EURASIP Journal of Applied Signal Processing*, No.3, pp. 382–396, 2005.

[11] M. Z. Win and R. A. Scholtz, "Impulse radio: how it works?," *IEEE Communication Letter* Vol. 2, pp. 36–38, Feb. 1998.

[12] T.H. Lee, *The design of CMOS radio-frequency integrated circuits*, 2nd ed. Cambridge University Press, 2004.

[13] D. D. Wentzloff and A. P. Chandrakasan, "Gaussian pulse generators for subbanded ultra-wideband transmitters," *IEEE Trans. Microwave Theory Tech.*, Vol. 54, pp. 1647–1655, June 2006.

[14] M. Reiha, J. R. Long, "A 1.2 V reactive-feedback 3.1–10.6 GHz low-noise amplifier in 0.13 μm CMOS," *IEEE Journal of Solid-State Circuits*, pp. 1023–1033, Vol. 42, No. 5, May, 2007.

[15] A. Tanaka, H. Okada, H. Kodama and H. Ishikawa, "A 1.1 V 3.1-to-9.5 GHz MB-OFDM UWB transceiver in 90 nm CMOS," in *IEEE ISSCC Dig. Tech. Papers*, Feb. 2006, pp. 120–121.

[16] A. Ismail, A. A. Abidi, "A 3.1 to 8.2 GHz zero-IF receiver and direct frequency synthesizer in 0.18 um SiGe BiCMOS for mode-2 MB-OFDM UWB communication," *IEEE Journal of Solid-State Circuits*, Vol. 40, No. 12, pp. 2573–2582, Dec. 2005.

[17] J. F. M. Gerrits, J. R. Farserotu, J. R. Long, "Multiple-user capabilities of FM-UWB communications systems," *Proc. International Conference on Utrawideband*, Sept. 2005, pp. 684–689.

[18] H. Gustat, F. Herzel, "Integrated FSK demodulator with very high sensitivity," *IEEE Journal of Solid-state circuits,* Vol. 38, No. 2, pp. 357–360, Feb. 2003.

[19] B. Zhou, H. Lv, M. Wang, et al., "A 1 Mb/s 3.2-4.4 GHz reconfigurable FM-UWB transmitter in 0.18 μm CMOS," *Proc. IEEE RFIC Symposium*, June, 2011, pp. 1–4.

[20] J. F. M. Gerrits, J. R. Farserotu, J. R. Long, "Multipath behavior of FM-UWB signals," *Proc. International Conference on Utrawideband*, Sept. 2007, pp. 162–167.

[21] E. Antoniono-Daviu, M. Cabedo-Fabrés, M. Ferrando-Bataller, and A. Vila-Jimenez, "Active UWB antenna with tunable band-notched behaviour," *Electronics Letters*, Vol. 43, No. 18, pp. 959–960, 2007.

[22] B. Razavi, *RF Microelectronics*, Prentice-Hall, first ed., 1998.

[23] B. Jagannathan, R. Groves, D. Goren, et al., "RF CMOS for microwave and mm-wave applications," *Proc. of Silicon Monolithic Integrated Circuits in RF Systems*, Jan. 2006, pp. 259–264.

[24] The MOSIS service, available on: http://www.mosis.com

[25] N. Saputra, J. R. Long, "A fully-integrated, short-range, low data rate FM-UWB transmitter in 90 nm CMOS," *IEEE Journal of Solid State Circuits*, Vol. 46, No. 7, pp. 1627–1635, July 2011.

[26] C. Mishra, A. V. Garcia, F. Bahmani, A. Batra, E. S. Sinencio, J. S. Martinez, "Frequency planning and synthesizer architectures for multiband OFDM UWB radios," *IEEE Transaction on Microwave Theory and Techniques*, Vol. 53, No. 12, pp. 3744–3756, Dec. 2005.

[27] R. B. Staszweski, S. Vemulapalli, P. Vallur, J. Wallberg, P. T. Balsara, "1.3V 20ps time-to-digital converter for frequency synthesis in 90-nm CMOS," *IEEE Trans.on Circuit and Systems II*, Vol. 53, No.3, pp. 220–224, 2006.

[28] Y. Liu, K. Contractor, Y. Kang, "Path loss for short range telemetry," *Proc. on 4th International workshop on wearable and implantable body sensor networks*, Vol. 13, 2007, pp. 70–74.

[29] M. Chiani and A. Giorgetti, "Coexistence between UWB and narrow-band wireless communication systems," *Proceedings of the IEEE*, Vol. 97, No. 2, pp. 231–254, 2009.

[30] J. F. M. Gerrits, H. Bonakdar, M. Detratti, et al., "A 7.2 −7.7 GHz FM-UWB transceiver prototype," *IEEE International conference on Ultra-wideband*, Sept., 2009, pp. 580–585.

[31] B. Zhou, R. He, J. Qiao, et al., "A low data rate FM-UWB transmitter with $\Delta-\Sigma$ based subcarrier modulation and quasi-continuous frequency-locked loop," in *Proc. IEEE A-SSCC*, Nov. 2010, pp. 33–36.

[32] S. Di Pasoli, "Fundamental limits to power consumption of LC sub-threshold oscillator," *Electronics Letters*, Vol. 44 No. 1, pp. 13–14, Jan. 2008.

[33] P. Nilsson, J. Gerrits, and J. Yuan, "A low complexity DDS IC for FM-UWB applications," in *Proc. 16th IST Mobile & Wireless Communications Summit*, July 2007.

[34] Y. Dong, Y. Zhao, J. F. M. Gerrits, G. van Veenendaal, J. R. Long, "A 9mW high band FM-UWB receiver front-end," *Proceeding of ESSCIRC*, Sep. 2008, pp. 302–305.

[35] J. F. M. Gerrits, J. R. Farserotu, J. R. Long, "A wideband FM demodulator for a low-complexity FM-UWB receiver," *Proceeding of the 9th European Conference on Wireless Technology*, Sept. 2006.

[36] K. Bult, "The effect of technology scaling on power dissipation in analog circuits", *Analog Circuits Design, Springer*, pp. 251–294, 2006.

[37] D. M. Binkley, B. J. Blalock, and J. M. Rochelle, "Optimizing drain current, inversion level, and channel length in analog CMOS design," *Journal of Analog Integrated Circuits and Signal Processing*, Vol. 47, pp. 137–163, April 2006.

[38] R. Hogervorst, S. M. Safai, J. P. Tero, and J. H. Huijsing, "A programmable 3-V CMOS rail-to-rail opamp with gain boosting for driving heavy loads," in *Proc. IEEE Int. Symp. Circuits Syst.*, 1995, pp. 1544–1547.

[39] C. DeVries and R. Mason, "A 0.18um CMOS, high Q-enhanced bandpass filter with direct digital tuning," *proc. of IEEE Custom Integrated Circuits Conference*, May 2002, pp. 279–282.

[40] A. Molnar, B. Lu, S. Lanzisera, B. W. Cook, and K. S. J. Pister, "An ultra-low power 900 MHz RF transceiver for wireless sensor networks," in *Proc. IEEE Custom Integrated Circuits Conf.*, 2004, pp. 401–404.

[41] B. W. Cook, A. Berny, A. Molnar, S. Lanzisera, K. S. J. Pister, "Low-power 2.4-GHz transceiver with passive RX front-End and 400-mV Supply," *IEEE journal of solid state circuits*, Vol. 41, No. 12, pp. 2757–2766, Dec. 2006.

[42] Q. Shi, "Power management in networked sensor radio a network energy mode," *IEEE Sensors Applications Symposium*, Feb. 2007, pp. 1–5.

[43] L. Stoica, A. Rabbachin, I. Oppermann, "A low-complexity noncoherent IR-UWB transceiver architecture with TOA estimation," *IEEE Transaction on Microwave Theory and Techniques*, Vol. 54, No. 4, pp. 1637–1646, April 2006.

[44] B. Peng, G. Huang, H. Li, P. Wan, P. Lin, "A 48-mW, 12-bit, 150-MS/s pipelined ADC with digital calibration in 65nm CMOS," *Proc. Custom Integrated Circuits Conference*, Sept. 2011, pp. 1–4.

[45] F. Sebastiano, L. J. Breems, K. A. A. Makinwa, S. Drago, D. M. W. Leenaerts, B. Nauta, "A 65-nm CMOS temperature-compensated mobility-based frequency reference for wireless sensor networks," *IEEE Journal of Solid-State Circuits*, Vol. 40, No. 7, July 2011, pp. 1544–1552.

[46] D. M. Binkley, *Tradeoff and optimization in analog CMOS design*, John Wiley and Sons Limited, 2008.

[47] J. N. Burghartz, W. C. Finley, J. J. Bastek, S. Moinian, I. A. Koullias, "High Q inductors for wireless applications in a complementary silicon bipolar process," *Proc. Bipolar and BICOMS Circuit and Tech. Meeting*, 1994, pp. 179–182.

[48] E. H. Armstrong, "A method of reducing disturbances in radio signaling by a system of frequency modulation," *Proceedings of the IRE* 24 Vol.5, pp. 689–740, May 1936.

[49] E. W. Herold, "Negative resistance and devices obtaining it", *Proc. IRE*, Vol. 23, No. 10, pp. 1201–1223, Oct. 1935.

[50] H. W. Bode, *Network analysis and feedback amplifier design*, New York: van Nostrand, 1945.

[51] N. Stanic, A. Balankutty, P. R. Kinget, Y. Tsividis, "A 2.4 GHz ISM band sliding-IF receiver with a 0.5-V supply," *IEEE Journal of Solid-State Circuits*, Vol. 43, No. 5, pp. 1138–1145, May 2008.

[52] S. Chatterjee, Y. Tsividis, P. Kinget, "0.5-V analog circuit techniques and their application in OTA and filter design," *IEEE Journal of Solid-State Circuits*, Vol. 40, No. 12, pp. 2373–2387, Dec. 2005.

[53] D. C. Daly, A. P. Chandrakasan, "An energy efficient OOK transceiver for wireless sensor networks," *IEEE Journal of Solid State Circuits*, Vol. 42, No. 5, pp. 1003–1011, May 2007.

[54] E. Y. Lin, J. M. Rabaey, A. Wolisz, "Power efficient rendezvous schemes for dense wireless sensor networks," *IEEE International Conference on Communications*, June 2004, pp. 3769–3776.

[55] F. Sebastiano, S. Drago, L. Breems, D. Leenaerts, K. Makinwa, B. Nauta, "Impulse based scheme for crystal-less ULP radios," *IEEE Transaction on Circuits and Systems-I*, Vol. 56, No. 5, pp. 1041–1052, May 2009.

[56] A. Tasic, S. T. Lim, W. A. Serdijn, J. R. Long, "Design of multimode RF front-end circuits," *IEEE Journal of Solid State Circuits*, Vol. 42, No. 2, pp. 313–322, Feb., 2007.

[57] M. W. Rashid, A. Garimella, P. M. Furth, "Adaptive biasing technique to convert pseudo-class AB amplifier to class AB," *Electronics Letters*, Vol. 46, No. 12, pp. 820–822, June 2010.

[58] B. W. Cook, A. Molnar, K. S. J. Pister, "Low power RF design for sensor networks", *RFIC symposium*, 2005, pp. 1–4.
[59] M. C. M. Soer, E. A. M. Klumperink, Z. Ru, F. E. van Vliet, B. Nauta, "A 0.2-to-2.0GHz 65nm CMOS receiver without LNA achieving $>$11 dBm IIP3 and $<$6.5dB NF," *Proc. of International Solid State Circuits Conference*, Feb. 2009, pp. 221–223.
[60] A. Vouilloz, M. Declerq, C. Dehollain, "A low-power CMOS super-regenerative receiver at 1GHz," *IEEE Journal of Solid State Circuits*, Vol. 36, No. 3, pp. 440–451, March, 2001.
[61] Y. Zheng, Y. Tong, C. W. Ang, Y. P. Xu, W. G. Yeoh, "A CMOS carrier-less UWB transceiver for WPAN applications," *ISSCC Dig. Tech, Papers*, Feb. 2006, pp. 116–117.
[62] S. Haykin, *Communication system 4th Ed.*, John Wiley & Sons, 2001.

3

FM-UWB Transmitter

3.1 Introduction

The RF transmitter contributes a significant portion of the total power consumed in an FM-UWB wireless transceiver system. The RF blocks, namely RF oscillator and PA, usually consume the most power in the transmitter chain (up to 80%), and therefore should be designed to be as power-efficient as possible. Fortunately, using FM-UWB relaxes the phase noise requirement on the transmit oscillator [1, 2] (see Section 2.5). With no synchronization or PLL synthesizer required and relaxed PA linearity constraints, a simple and low-power RF transmitter prototype can be built, where most of the building blocks are fully integrated on-chip.

Based on the specifications listed in Table 2.1, a ring oscillator is chosen to generate the carrier. It is followed by a power-optimized transmit amplifier instead of a simple buffer. Additionally, a calibration circuitry is used to compensate for process, supply voltage, and temperature (PVT) variation. Baseband and bias circuits are included on the chip which ease characterization.

The FM-UWB transmitter building blocks and their design are described in Section 3.2 of this chapter. Measurement results of a prototype realized in 90-nm bulk CMOS are presented and discussed in Section 3.3. A comparison with other low-power FM-UWB transmitters from the recent literature is presented in Section 3.4.

3.2 FM-UWB Transmitter Circuit Designs

The proposed fully-integrated FM-UWB transmitter is shown in Figure 3.1. A current-controlled ring oscillator (ICO) is the source for the RF carrier. The transmit signal is buffered and amplified by a multi-stage power amplifier that drives the antenna through an external matching network. The triangular

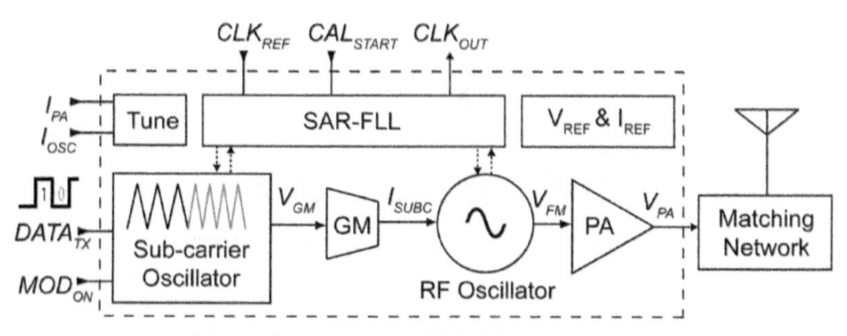

Figure 3.1 Proposed FM-UWB transmitter.

sub-carrier is generated using a relaxation oscillator, where the triangular oscillator output voltage is converted to a current by a transconductor, GM. An on-chip successive approximation register frequency-locked loop (SAR-FLL) calibrates the sub-carrier and RF oscillator frequencies. Voltage and current references are also implemented on-chip to generate supply and temperature insensitive references and for general biasing. External (analog) currents can also be injected for testing through CMOS current mirrors (i.e., I_{OSC} for tuning the ICO output frequency, and I_{PA} for changing the PA output power).

The prototype FM-UWB transmitter is implemented in 90-nm bulk RF-CMOS with thick metal and metal-insulator-metal (MIM) capacitor options [3]. The all-copper interconnect scheme (from bottom to top of the stack) consists of 5 thin, two medium-thick, and one thick top metal.

3.2.1 RF Carrier Oscillator

Well-known for its simplicity, compact physical layout, and wide tuning range, a ring topology is chosen for the RF transmit oscillator. The ring oscillator benefits from technology scaling because its power consumption decreases as the supply voltage and transistor sizes are reduced. Three stages are used to meet the desired oscillation frequency with minimal power consumption, but at the cost of higher phase noise [4]. As mentioned before in the specifications (Section 2.5), the phase noise requirement of the RF oscillator is relaxed, therefore this topology is a suitable choice. Tuning of the control current is independent of the supply voltage, making the current-controlled oscillator (ICO) less susceptible to noise present on the supply or ground lines, these aspects of the design improve its portability to advanced technology nodes, where further reduction of the supply voltage due to scaling is anticipated.

Each stage of the ICO consists of an NMOS driver with active PMOS load, as shown in Figure 3.2. Regulation of the ring oscillator supply desensitizes it

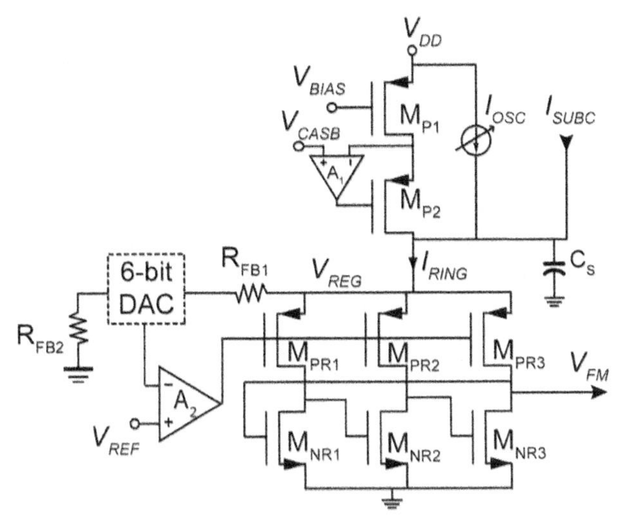

Figure 3.2 Current-controlled RF oscillator (RF-ICO) schematic.

to variations in the supply voltage, V_{DD}. The PMOS ($M_{PR1\text{-}3}$) source terminal voltage (V_{REG}) is regulated at 0.7 V by opamp A_2 via a feedback loop that derives the desired supply-voltage using temperature-compensated voltage reference, V_{REF}. The opamp bandwidth must be at least ten times larger than the sub-carrier frequency for proper regulation of the ICO supply. The ICO output frequency (f_{osc}) is tuned by current I_{RING}, which is the sum of the bias current from M_{P1}, the external tuning current I_{OSC}, and the sub-carrier modulation current I_{SUBC}, all of which are supplied by PMOS current sources. The current sources are implemented in parallel from a high ratio (>100) current mirror, hence their output impedance is low. Feedback amplifier A_1 enhances the current sources' output impedance, improves current mirror accuracy, and yields 65-dB power supply rejection from DC up to 10 kHz.

The ICO is based on a current-starved design [5], where the rise/fall time of each stage is slew rate limited. The output frequency (f_{osc}) for the 3-stage ring oscillator may be estimated from

$$f_{OSC} = \frac{I_{RING}}{6C_G|V_{FM}|}, \tag{3.1}$$

where C_G is the total capacitance seen at all the NMOS gates (including Miller effect) and $|V_{FM}|$ is the output voltage swing. Equation (3.1) predicts that I_{RING} is linearly proportional to f_{osc} for constant $|V_{FM}|$, however, there is non-linearity caused by slight changes in $|V_{FM}|$. The voltage swing decreases slightly as the frequency increases (see Figure 3.3a), because the loop gain decreases with increasing frequency so that the output swing settles at a lower peak-to-peak voltage. Another advantage of tying f_{osc} to the current reference is that it can be made temperature insensitive. Simulation result shown in Figure 3.3b indicates that the ICO temperature sensitivity is 1.25 MHz/°C.

Aspect ratios (W/L) of the transistors in each stage of the ICO are scaled progressively by a factor of two to increase the current driving capability at output, V_{FM}. Scaling also balances the overall capacitive loading on each stage, which helps realize a higher oscillation frequency. RF carrier power increases from the first to third stage, while the voltage swing in each stage remains constant. The 1-pF decoupling capacitor (C_S), as shown in Figure 3.2,

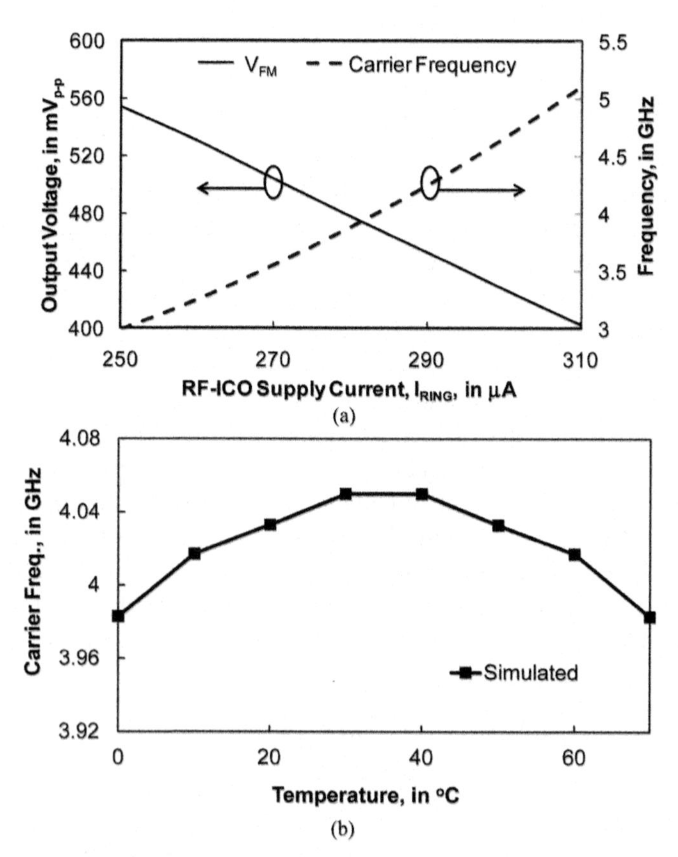

Figure 3.3 Simulated (a) frequency and amplitude of the RF-ICO versus current bias, and (b) frequency versus temperature.

acts as a charge reservoir that reduces fluctuations in the node voltage V_{REG}, and also filters out high-frequency supply noise. A 6-bit resistor-DAC is used to tune the ICO frequency during calibration by adjusting V_{REG}.

The ICO (incl. opamps and bias circuits) consumes 280 μA of DC current when it is oscillating at 4 GHz. A change in the tuning current of 15 μA causes 500-MHz deviation in the RF output, resulting in a tuning sensitivity of 33 MHz/μA.

3.2.2 Power Amplifier

The largest proportion of the total transmitter power should be dedicated to the power amplifier (PA) to maximize the overall transmitter power efficiency. FM-UWB produces a constant envelope RF signal that enables the PA to operate at its maximum power output where its efficiency is highest. The output power of the transmitter is adjusted via the bias current I_{PA} (see Figure 3.4), which controls the bias of the last stage. Separate bias voltages reduce coupling between stages and avoids self-oscillation in the multi-stage amplifier (biasing circuits are not shown).

The PA consists of 3 push-pull stages in cascade as shown in Figure 3.4. Class-AB biasing increases the power efficiency of the PA by drawing peak current only during the charging phase. This comes at the cost of reduced linearity, which is less important as the carrier amplitude is approximately constant for FM-UWB. Additionally, the PA provides isolation between the antenna and the ICO, as a strong interferer at the transmitter output could cause unwanted pulling of the ICO frequency. The simulated reverse isolation

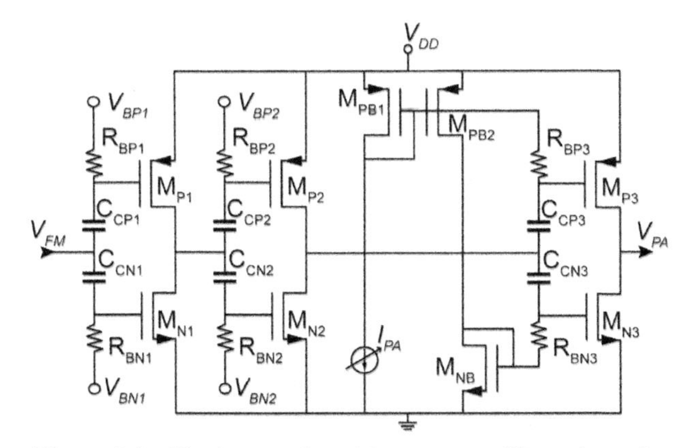

Figure 3.4 The 3-stage class-AB power amplifier schematic.

from transmitter to ICO outputs is −61 dB below 10 GHz, although this result must be considered optimistic because parasitics from interconnect lines are not included in the simulation.

The RF signal is coupled from stage to stage through R-C bias-T networks (R_{Bx} and C_{Cx} in Figure 3.4). The multi-stage amplifier is scaled by 2.5 times from stage to stage, and the PMOS to NMOS scaling factor is 1.5. The gain per stage is 5 dB, and the (simulated) input power to the PA from the oscillator is −25 dBm, which is sufficient to drive the output stage into saturation (0.85-$V_{p\text{-}p}$ drain voltage) for maximum efficiency at targeted peak power of 100 μW. A matching network transforms the 50-Ω load impedance to the optimum load of 900 Ω for the amplifier output.

The PA consumes 750 μW when running at 4 GHz, and produces −10 dBm (maximum) output power into a 50-Ω load via the external LC matching network, which consists of a 6-nH shunt inductor and 0.3-pF series capacitor. The simulated peak power-added efficiency (PAE) of the PA is 12.9%.

3.2.3 Sub-Carrier Oscillator

The 0.6 to 1.2 MHz triangular sub-carrier signal is generated by the relaxation oscillator shown in Figure 3.5. Current reference I_{REF} charges and discharges capacitor C_{SUB}, thereby generating a ramp voltage (V_{TRI}) across the capacitor. The peak-to-peak amplitude of the ramp is determined by the upper (V_{REFH})

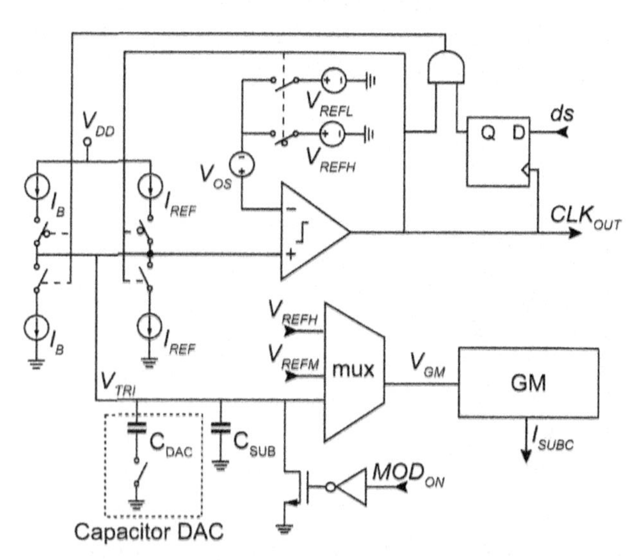

Figure 3.5 Sub-carrier oscillator schematic.

and lower (V_{REFL}) trigger points of the comparator, which determine the switching of the capacitor current between charge and discharge modes. The voltage across C_{SUB} (V_{TRI}) and the comparator output are set to 0 V initially to ensure proper startup of the triangle generator when the modulation is off. The input data stream (ds) modulates the frequency of the sub-carrier oscillator by injecting an additional current I_B (25% of I_{REF}) in parallel with I_{REF} into capacitor C_{SUB} when the input data is 'high'. A D-flip-flop synchronizes frequency shifts of the sub-carrier to occur only when the ramp generator toggles between charge and discharge modes.

The sub-carrier frequency (f_{SUB}) generated by the relaxation oscillator is given by

$$f_{SUB} = \frac{I_{REF} + (ds.I_B)}{2C_{SUB}(V_{REFH} - V_{REFL})}. \tag{3.2}$$

A linear (MIM) rather than a MOS capacitor is used to realize C_{SUB} to improve the linearity of the ramp generator output. There are other circuit imperfections that affect f_{SUB}, including channel length modulation of MOS current sources, variation in propagation delays of the comparator and control logic circuits, and variation in the value of C_{SUB}. Thus, a 6-bit capacitor-DAC in parallel with C_{SUB} is added to adjust the ramp capacitor value and tune the sub-carrier frequency with a resolution of 3.5 kHz, or within 0.2% of the desired sub-carrier frequency. As mentioned before, FM-UWB is non-coherent and only requires frequency accuracy on the order of 10^3 ppm (e.g., 5000 ppm from [5]), or about one order of magnitude poorer than the stability required to track the carrier phase in a coherent system.

Current I_{REF} and reference voltages V_{REFL} and V_{REFH} are derived from the same source to desensitize the circuit from PVT variations. This topology is also insensitive to the comparator offset (V_{OS}). Offset affects both upper and lower threshold voltages equally, therefore any offsets in the comparator reference voltages are canceled as seen from Equation (3.2). The amplifier input is initialized to V_{REFM} (mid-level between V_{REFH} and V_{REFL}) through a multiplexer (mux) when the modulation is turned off (i.e., $\text{MOD}_{\text{ON}} = 0$), allowing the RF-ICO to be calibrated at its desired center frequency. The voltage corresponding to maximum RF frequency deviation, V_{REFH} can be selected through the mux to accommodate RF bandwidth calibration.

In a revised version of the transceiver design (see Chapter 5), a 9-bit capacitor DAC is implemented to extend the frequency range to 0.5–4 MHz. Also, an option to boost I_B to 50% of I_{REF} is included to increase the FSK frequency deviation, which permits a higher data rate.

3.2.4 Transconductance Amplifier

Transconductance amplifier A_0 translates the sub-carrier voltage to a current that modulates the transmit RF-ICO via feedback resistor R_{GM} (see Figure 3.6). R_{GM} is external to the transceiver chip, and is tunable to control the bandwidth of the wideband frequency modulation. Unfortunately, I_{SUBC} consists of a DC offset plus modulating current. If R_{GM} is tuned to increase bandwidth, the DC offset also increases and shifts the center frequency of the ICO. A better solution requires a sink-source type of current output without a DC offset, which is implemented in the transceiver described in Chapter 5.

Amplifier A_0 is a 2-stage, folded-cascode topology with 70-dB DC gain and 20-MHz unity-gain bandwidth. Wide bandwidth is required to reproduce the triangle wave with the desired fidelity (i.e., harmonics preserved). The output current source formed by transistors M_{P1} and M_{P2} is gain boosted by the amplifier A_1 to improve the accuracy of the mirrored current and to enhance its low-frequency output impedance.

3.2.5 SAR-FLL Calibration

Simulations predict that the sub-carrier and RF-ICO center frequencies vary by up to 10% due to PVT spreads, and hence require calibration. The transmitter carrier and sub-carrier frequencies are calibrated to 0.5% and 0.2% accuracy, respectively. A block diagram of the digital frequency-locked loop (FLL) used to set the initial carrier frequency is shown in Figure 3.7. It consists of a digital

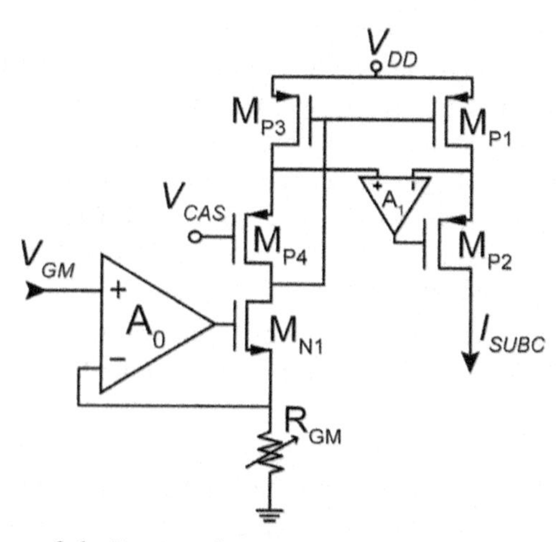

Figure 3.6 Transconductance amplifier (GM) schematic.

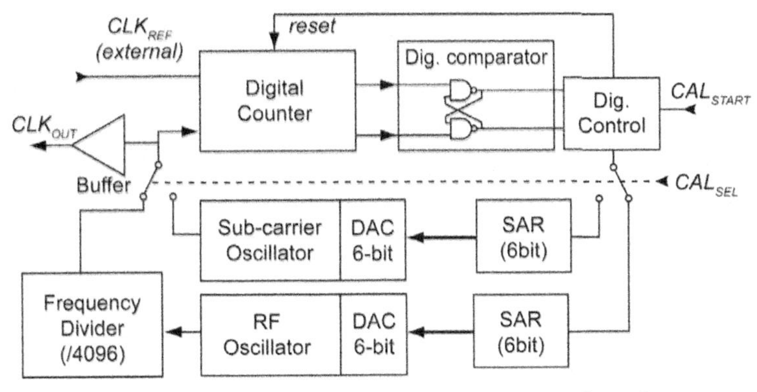

Figure 3.7 The SAR frequency-locked loop calibration scheme.

controller, a 12-stage frequency divider, a digital counter, a 6-bit successive approximation register (SAR), and a 6-bit DAC. A successive binary search algorithm is implemented that ensures convergence of the frequency tuning loop.

The calibration sequence is initiated when a pulse is applied to the positive-edge-triggered CAL_{START} input (at t_1) and the controller (Dig. Control) resets the digital counter (at t_2), as shown in Figure 3.8. The digital counter integrates both the divide-by-2^{12} RF-ICO output (CLK_{OUT}) and the external reference clock (at t_3). A digital comparator senses the frequency difference (at t_4), and adjusts the RF-ICO frequency via the SAR and the DAC. N cycles of successive approximation (where N is the SAR resolution; $N = 6$ in this case) tunes the divided output of the RF-ICO to the external reference clock frequency. Note that the SAR cycles are not synchronous; a new SAR cycle starts when any of the digital counters reach a certain

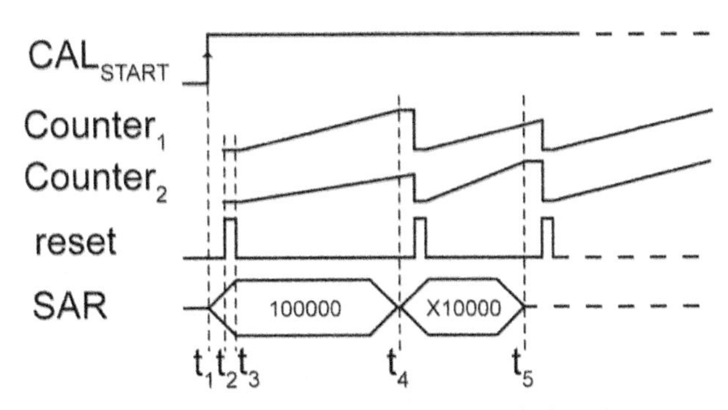

Figure 3.8 SAR-frequency-locked loop timing diagram.

threshold. The counter threshold could also be made digitally programmable, if desired. To provide a programmable frequency ratio, a variable-frequency external reference clock is assumed. The external reference clock would be sourced from the baseband digital processor in a wireless sensor application. Calibration requires 6 ms (i.e., 1 ms/SAR cycle) for a nominal 1-MHz external clock. The same calibration loop is also used to tune the sub-carrier oscillator by selecting the appropriate oscillator output and SAR input using the CAL_{SEL} switch.

The calibration setting is stored until another calibration sequence is initiated, which may be necessary if there are changes in the supply voltage or ambient temperature. Simulation predicts that the calibrated frequency accuracy is within 1.5% when operating at supply voltage of 0.9-1.1 V, and within 2.5% when operating at a temperature of 0°–80°C.

The RF divider is implemented using a cascade of 12 D-flip-flops, as shown in Figure 3.9a. The first 6-stages are implemented using a true, single-phase clock (TSPC) flip-flop [7] shown in Figure 3.9b, while the remainders are conventional CMOS pass-gate flip-flops [8]. Using only a few transistors and precharged nodes, the TSPC flip-flop is suitable for dividing a high frequency (above 50 MHz), periodic (i.e., clock) signal. At lower frequencies, charge leakage from the precharged nodes affects reliability of the divider output. The frequency divider is active only during the calibration process, and it consumes 35 μW when running continuously.

The SAR is controlled by a state machine as shown in Figure 3.10. The bits that are stored in the SAR are the state address. When calibration begins, the SAR address is initialized with an MSB trial. The MSB is kept at logic '1' or changed back to logic '0' depending on the digital comparator result (*Comp*). The state address is changed progressively from MSB to LSB. A 6-bit shift register generates a traveling pulse which marks which bit need to be changed. It is initialized with the CAL_{START} signal and produces a CAL_{DONE} output signal when the calibration cycle is finished. The state address is changed by arithmetically adding (or subtracting) the current address to the marker address from the shift register. Subtraction is implemented using adder circuitry and two's complement arithmetic. A multiplexer is employed to select between the non-inverted or inverted bits, which effectively decides the sign of the result based on the *Comp* state. An Asynchronous calibration clock (CAL_{CLK}), generated when each SAR bit is decided, drives the state machine. Using an asynchronous clock speeds up the calibration process [9].

The SAR-FLL scheme employed here does not require precision components or high-speed circuitry. The measured sub-carrier FLL locking range

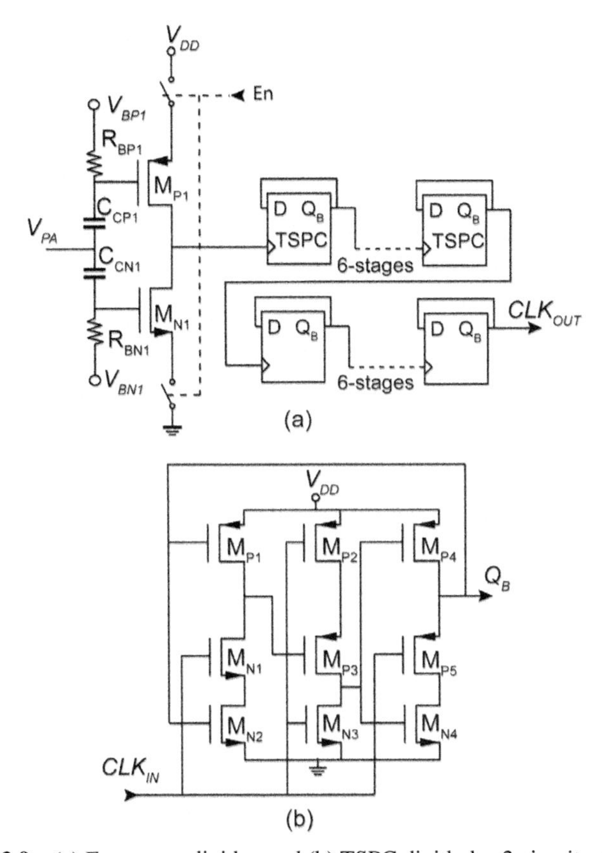

Figure 3.9 (a) Frequency divider, and (b) TSPC divide-by-2 circuit schematic.

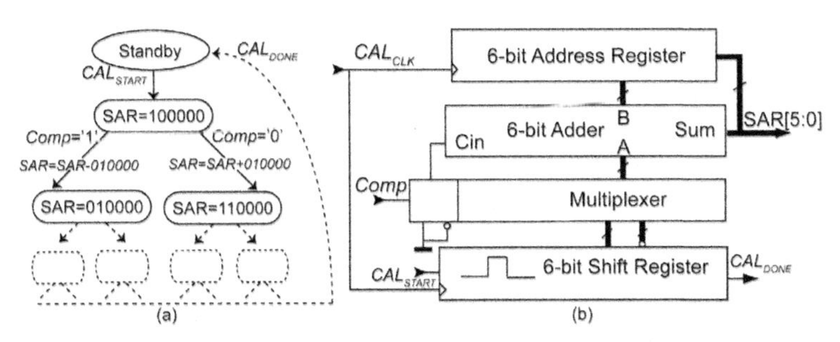

Figure 3.10 SAR state (a) and block (b) diagram.

is 580–750 kHz and 710–910 kHz for logic '0' and logic '1' input data, respectively. The carrier frequency generated by the RF-ICO can be locked between 3.4–4.8 GHz by the FLL with a resolution of 20 MHz, which is sufficient for communication between the transmitter and the receiver using the non-coherent FM-UWB scheme. In an improved version of the transceiver (see Chapter 5), the SAR-FLL also calibrates the RF bandwidth, ICO tuning range, and center frequency of the receiver.

3.2.6 Voltage and Current References

A circuit which supplies current and voltage biases and references is essential for highly-integrated systems on a chip, like the FM-UWB transmitter. The current and voltage references have to be well-controlled despite variations in the supply voltage, IC temperature, and processing. The circuit consists of a proportional-to-absolute-temperature (PTAT) current generator and an inverse-PTAT current generator as shown in Figure 3.11. The PTAT current is generated by applying the difference in the threshold voltage between different size NMOS transistors across polyresistor R_1, in a manner analogous to a conventional bandgap reference [10]. The inverse-PTAT is generated using an NMOS transistor's threshold voltage (with a negative temperature coefficient of approximately -1.5 mV/°C) dropped across polyresistor R_2. The reference current is given by

$$I_{REF} = \frac{V_{GS}}{R_2} + \frac{\Delta V_{GS}}{R_1}. \tag{3.3}$$

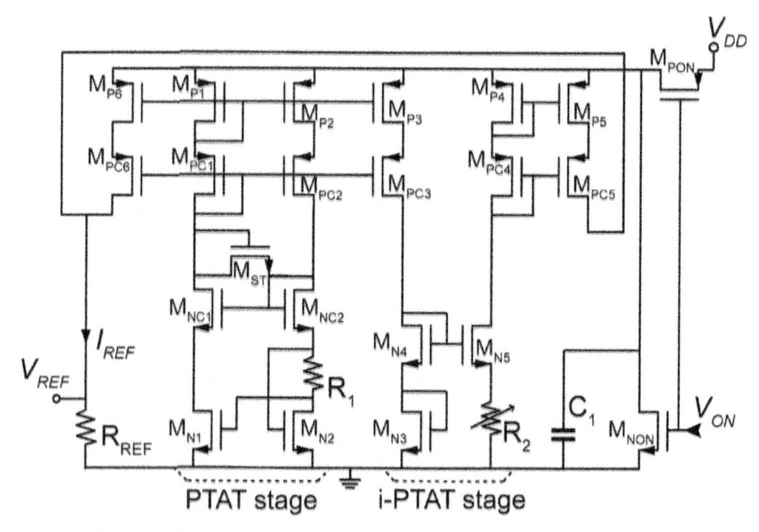

Figure 3.11 Schematic of built-in voltage and current reference.

Voltage reference (V_{REF}) is derived from I_{REF} flowing through polyresistor R_{REF}. Note that any variation in the polyresistor value on the reference current value is canceled as long as the resistors are matched (n.b., simulation models estimate 0.3% mismatch).

Resistor R_1 provides feedback to reduce the sensitivity of the current reference (I_{REF}) to supply voltage changes when operating at a 1-V supply. Capacitor C_1 filters any noise from the supply voltage V_{DD}. Diode connected M_{ST} ensures that the biasing loop operates during startup, and automatically shuts off when the correct operating point is reached.

The simulated current variation due to $\pm 10\%$ changes in supply voltage is 1.5%. The simulated current variation due to changes in temperature is approximately 200 ppm/°C at the nominal supply voltage of 1 V. These variations can be corrected (if required) by trimming or further calibration. In the updated version of transceiver (see Chapter 5), R_2 is trimmed by 2-bit switches to minimize current variation due to process variation. The reference circuit can be turned on/off by V_{ON} to reduce current consumption during times when the transmitter is inactive.

3.3 Experimental Measurement

The die photo of the transmitter prototype is shown in Figure 3.12. Note that the testchip is pad-limited, as some I/Os are used only for diagnostic/verification purposes. Area not required by the active circuitry is occupied by supply decoupling capacitors. Each pad is ESD protected using a double-diode cell that offers human body model (HBM) protection of 2 kV. On-chip inductors are not used in the design, giving a relatively small active area of just 0.1 mm^2

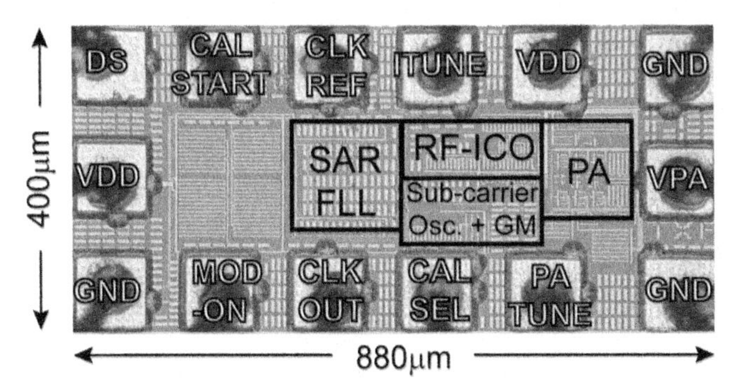

Figure 3.12 Chip micrograph of the FM-UWB transmitter.

(excluding bondpads). All measurement results presented in this section are for a packaged sample, where the test die is wirebonded to a custom printed-circuit board (PCB) and test fixture for characterization, as shown in Figure 3.13. The laminate used for the circuit board has a loss tangent of 0.004 at 5 GHz. A 50-Ω transmission line fabricated on this material has a (measured) insertion loss of 0.07 dB/cm at 4 GHz. Transmission lines are used to connect the IC to SMA connectors at the board's periphery. The bondwire length between the RF input on the chip and the PCB is approximately 1 mm, resulting in a series inductance of approximately 1 nH. The supply lines are filtered using 10-nF and 330-pF decoupling capacitors in parallel on the PCB side, as well as a 50-pF MIM decoupling capacitor implemented on-chip.

The measured RF spectrum with a resolution bandwidth of 1 MHz is shown in Figure 3.14. The unmodulated carrier power in this measurement is −14 dBm, and the spectral bandwidth of the FM-UWB output, when modulated by the on-chip sub-carrier generator, is 500 MHz, with a peak power output of −41.4 dBm. The variation in RF output power across the FM-UWB spectrum is less than 2 dB. The maximum unmodulated output power is plotted versus frequency in Figure 3.15. However, there is some variation in the output power due to the matching network. Across 500-MHz bandwidth, the flatness in the response is within 1 dB, which is acceptable. The spectral

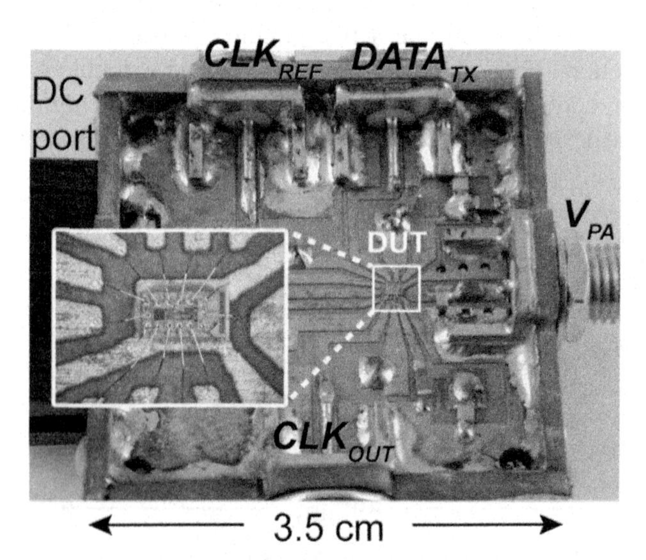

Figure 3.13 Assembled test PCB.

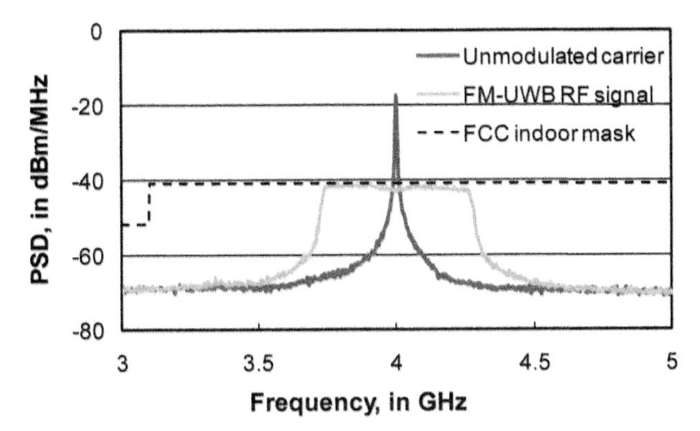

Figure 3.14 Measured transmitter output spectrum (unmodulated and modulated).

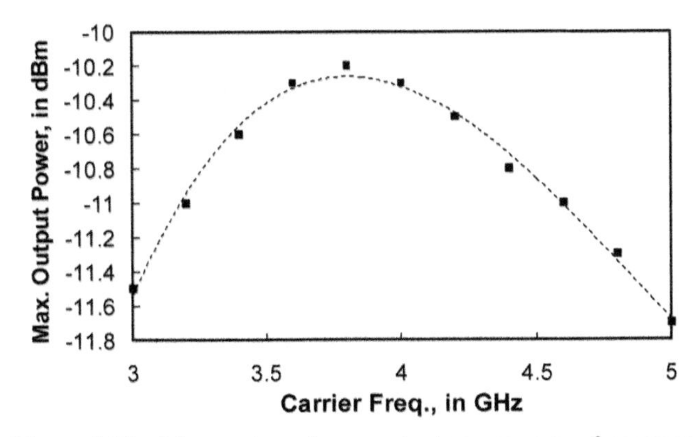

Figure 3.15 Measured maximum output power versus frequency.

flatness can be controlled by dynamically adjusting I_{PA} (see Figure 3.4) using a fraction of the sub-carrier current.

The measured and simulated sensitivity of the carrier frequency to changes in supply voltage are plotted in Figure 3.16. The data shows that measurement and simulation agree well, and that the current-controlled tuning input is relatively insensitive to changes between 1 V and 1.1 V in supply voltage. The relationship between the RF carrier frequency and tuning current is plotted in Figure 3.17. The measured behavior is close to linear, as predicted by Equation (3.1), but with a small non-linearity due to slight changes in the oscillation amplitude. This can cause non-uniformity in the power spectral density (PSD) seen at the output. Given that FM-UWB systems are more tolerant to multipath transmission effects (e.g., fading), which also affect

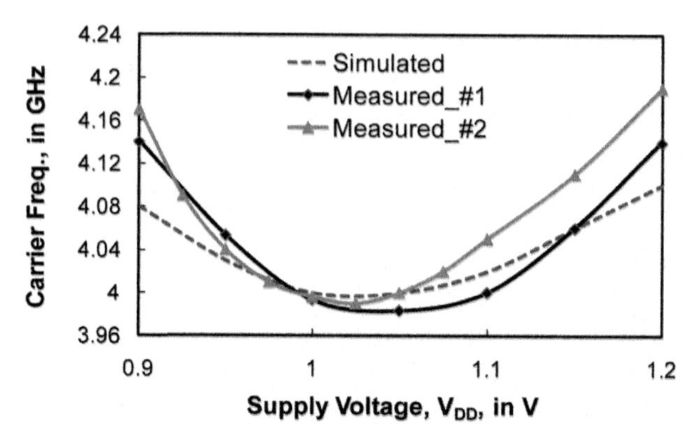

Figure 3.16 Carrier frequency vs. supply voltage, V_{DD}.

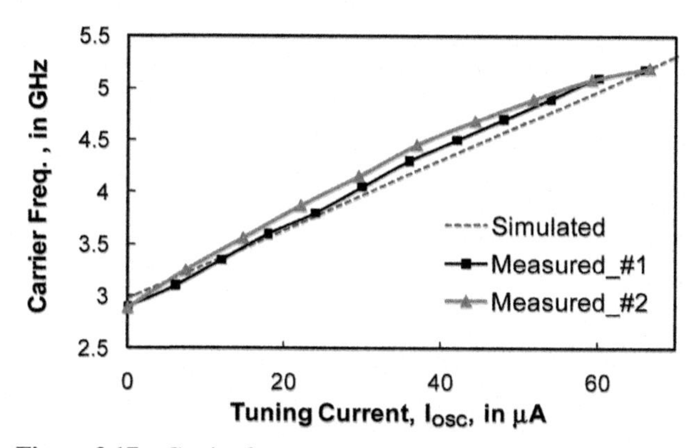

Figure 3.17 Carrier frequency vs. RF-ICO tuning current, I_{OSC}.

the uniformity of the transmit PSD, the non-linearity is not expected to cause significant impairments. However, the output power might have to be reduced slightly to satisfy the −41 dBm/MHz transmit power regulation. The maximum measured oscillation frequency of 5.2 GHz is limited by the output swing of amplifier A_2 (see Figure 3.5). Figure 3.18 shows results from phase noise measurement of the transmitter at 4-GHz center frequency and −12 dBm output power. The resulting phase noise is −75 dBc/Hz at 1 MHz offset, which exceeds the specification of −62 dBc/Hz. An advantage of the FM-UWB modulation scheme is that oscillator phase noise performance can be relaxed, which lowers the overall power consumption of the oscillator and transmitter as a whole.

The measured performance of the FM-UWB transmitter is summarized and compared to other FM-UWB front-ends from the literature in Table 3.1.

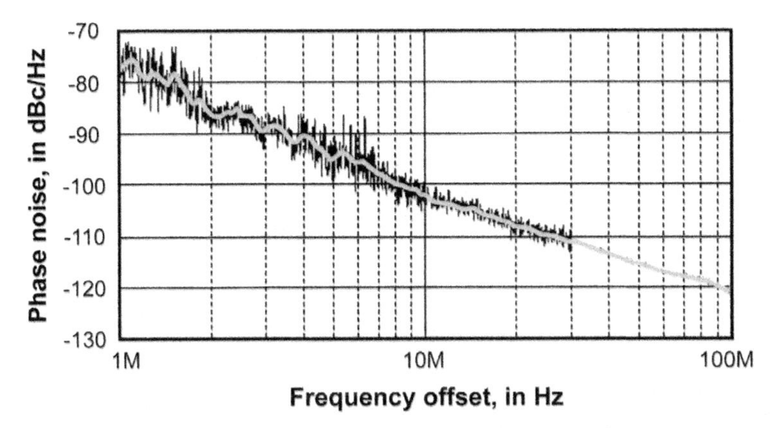

Figure 3.18 Measured phase noise of the transmitter.

Table 3.1 FM-UWB transmitter performance summary

Parameters	This Work	[11]	[12]	[13]
Technology	90 nm CMOS	180 nm CMOS	180 nm CMOS	130 nm CMOS
RF tuning range, in GHz	2.9–5.2	2.7–4.1	0.5–5	6.25–8.25
V_{DD}, in V	1	1.8	1.8	1.1
Phase noise, in dBc/Hz at 1 MHz	−75	−70	−75	−107
Max. output power, in dBm	−10.2	−34	−9	−5
Bandwidth, in MHz	500	–	550	550
Sub-carrier frequency, in MHz	0.8	–	1	1
Data rate, in kbit/s	100	100	100	100
Power consumption, in mW	0.8–1.1	7.2–14	2.5–10	4.6
Active area, in mm^2	0.1	0.7*	0.25*	0.062
Energy efficiency, in nJ/bit	8–11	72–140	70–100	46

*Including bondpads

The transmitter consumes one-quarter of the power, but has a much higher integration level when compared to other designs listed in the table. The energy consumption of the transmitter running continuously (i.e., not duty-cycled) at 4 GHz (center frequency) is 9 nJ/bit at a data rate of 100 kbit/s. The maximum overall power efficiency (RF output power divided by total DC power) of the full transmitter is 9.1%. Compared to the other FM-UWB transmitters, the IC developed in this work integrates all self-calibration, voltage and current reference circuits on-chip and consumes less power. Energy efficiency is also improved by approximately a factor of 5 compared to the 0.13-μm CMOS design from [13].

3.4 Summary and Conclusions

FM-UWB offers an efficient utilization of spectrum available for unlicensed use when the center frequency is well-controlled due to the shape of its transmit power spectral density. A fully-integrated FM-UWB transmitter for the 3–5 GHz band has been realized in a low-cost 90-nm bulk CMOS technology, which benchmarks the scheme for potential low-power, short-range applications. The 0.35-mm^2 transmitter IC demonstrator consumes just 900 μW from a 1-V supply when operating at 4-GHz RF center frequency. A simple, low-cost and effective SAR-FLL calibration method was implemented to tune the center frequency of the FM-UWB transmitter. The nominal energy consumption of the transmitter is 9 nJ/bit at 100-kbit/s data rate in continuous operation, which is suitable for low data rate, portable wireless applications powered from a battery.

Table 3.2 compares the FM-UWB transmitters with other low-power transmitters. Energy efficiency in J/bit is an indicator of an efficient design. However, applications such as data logging and monitoring require low kbit/s data rates that typically yield poor energy efficiency as it is inversely proportional to the data rate. Duty cycling could be applied to this FM-UWB transmitter (and other designs) to improve efficiency. Another performance indicator is output power efficiency, but it is also dependent on maximum transmitted power. Generally, BFSK transmitters [15, 16] offer higher transmitter efficiency using a tuned, narrowband output stage rather than a wideband (e.g., UWB) scheme which consumes more power. However, the narrowband FSK transceivers will require additional power to improve their robustness to multipath fading and to supply self-calibration and biasing circuits required

Table 3.2 Low-power transmitter performance comparison

Parameters	This Work	[14]	[15]	[16]
Technology	90 nm CMOS	180 nm CMOS	130 nm CMOS	180 nm CMOS
Modulation	FM-UWB	OOK	BFSK	BFSK
Frequency, in GHz	4	0.9	2.4	2.4
Max. output power, in dBm	−10.2	−11.4	−5	−5.2
Data rate, in kbit/s	100	1000	300	125
Power consumption, in mW	0.9	3.8	1.12	1.15
Active area, in mm^2	0.1	0.27*	–	0.273
Energy efficiency, in nJ/bit	9	3.8	2.3	9.2
Power efficiency, in %	9.1	2.1	30	26.8

*Including Rx circuit

on-chip to accommodate IC manufacturing, supply voltage and temperature variations (included in our prototype FM-UWB transmitter). The FM-UWB transmitter designed in this work offers competitive energy efficiency at low data rates despite the overhead in power consumption associated with continuous operation. With on-chip calibration and a small chip area, the measured results establish FM-UWB as a viable alternative for short-range, low data rate wireless applications.

The next chapter will discuss the FM-UWB receiver prototype. The receiver uses the input RF signal generated by the FM-UWB transmitter described in this chapter for characterization purposes. The functionality of both the transmitter and receiver are validated from experimented measurements.

References

[1] J. F. M. Gerrits, M. Kouwenhoven, P. van der Meer, J. R. Farserotu, and J. R. Long, "Principles and limitations of ultra-wideband FM communication systems," *EURASIP Journal on Applied Signal Processing*, pp. 382–396, 2005.

[2] N. Saputra, J. R. Long, J. J. Pekarik, "A 900 μW 3-5GHz transmitter for FM-UWB in 90 nm CMOS," in *Proc. ESSCIRC,* Sept. 2010, pp. 398–401.

[3] B. Jagannathan, R. Groves, D. Goren, et al., "RF CMOS for microwave and mm-wave applications", *Proc. of Silicon Monolithic Integrated Circuits in RF Systems*, Jan. 2006, pp. 259–264.

[4] T. C. Weigandt, B. Kim, and P. R. Gray, "Analysis of timing jitter in CMOS ring oscillators," in *Proc. IEEE ISCAS*, pp. 27–30, June 1994.

[5] D. A. Badillo, S. Kiaei, "Comparison of contemporary CMOS ring oscillators," in Proc. *IEEE Radio Frequency Integrated Circuits Symposium*, June 2004, pp. 281–284.

[6] A. P. Chandrakasan, F. S. Lee, D. D. Wentzloff, et al., "Low-power impulse UWB architectures and circuits," *Proc. of the IEEE* Vol. 97, No.2, pp. 332–352, Feb. 2009.

[7] J. Yuan and C. Svensson, "High-speed CMOS circuit technique," *IEEE Journal of Solid State Circuits*, Vol. 24, No. 1, pp. 62–70, Feb. 1989.

[8] J. P. Uyemura, *Introduction to VLSI circuits and system*, John Wiley & Sons, 2002.

[9] S. W. M. Chen and R. W. Brodersen, "A 6-bit 600-Ms/s 5.3 mW asynchronous ADC in 0.13-μm CMOS," *IEEE Journal of Solid-State Circuits*, Vol.41, pp. 2669–2680, Dec. 2006.

[10] G. C. M. Meijer, G. Wang, and F. Fruett, "Temperature sensors and voltage references implemented in CMOS technology," *IEEE Sensors Journal*, Vol. 1, No. 3, pp. 225–234, Oct. 2001.
[11] T. Tong, Z. Wenhua, J. Mikkelsen, T. Larsen, "A 0.18μm CMOS low power ring VCO with 1GHz tuning range for 3-5GHz FM-UWB applications," *Proc. of 10th IEEE International Conference on Communication Systems*, 2006, pp. 1–5.
[12] A. Georgiadis, M. Detratti, "A linear, low power, wideband CMOS VCO for FM-UWB applications," *Microwave and Optical Technology Letters*, Vol. 50, No. 7, pp. 1955–1958, July 2008.
[13] M. Detratti, E. Perez, J. F. M. Gerrits, M. Lobeira, "A 4.6 mW 6.25–8.25 GHz RF transmitter IC for FM-UWB applications," *Proc. of* the *ICUWB*, pp. 180–184, Sep. 2009.
[14] D. C. Daly, A. P. Chandrakasan, "An energy efficient OOK transceiver for wireless sensor networks," *IEEE Journal of Solid State Circuits*, Vol. 42, No. 5, pp. 1003–1011, May 2007.
[15] B. W. Cook, A. Berny, A. Molnar, S. Lanzisera, K. S. J. Pister, "Low-power 2.4-GHz transceiver with passive RX front-End and 400-mV Supply," *IEEE Journal of Solid State Circuits*, Vol. 41, No. 12, pp. 2757–2766, Dec. 2006.
[16] J. Ayers, N. Panitantum, K. Mayaram, T. S. Fiez, "A 2.4GHz wireless transceiver with 0.95nJ/b link energy for multi-hop battery free wireless sensor networks," *Symp. on VLSI Circuits Dig. Tech. Papers*, June 2010, pp. 29–30.

4

FM-UWB Receiver

4.1 Introduction

The objective of the work in this chapter is to realize a power efficient receiver utilizing the advantages of a regenerative amplifier. The RF signal must be amplified so that it can be detected by an envelope detector without limiting the sensitivity. The RF preamplifier typically consumes the largest portion (>80%) of power in the receiver that employs envelope detection [1]. The FM-UWB receiver prototype described in this chapter is a positive feedback (i.e., regenerative) amplifier that enhanced RF gain (see Section 2.7.2 for detail). The amplifier is followed by an envelope detector to recover the IF sub-carrier.

The power consumption of the receiver cannot be reduced arbitrarily due to the constraints imposed by operating frequency, gain-bandwidth product, and noise. Noise generated by the preamplifier limits the receiver sensitivity. The receiver sensitivity is (in general) proportional to transconductance (gm) of the preamplifier. As gm is also proportional to the bias current and is therefore directly related to power consumption, the sensitivity is inversely (not linearly) proportional to power consumption. The receiver is designed to meet the specifications listed in Table 2.1, while consuming as little power as possible.

The operating principles of the proposed FM-UWB receiver are described in Section 4.2. Circuit design details of the receiver prototype and its operation are discussed in Section 4.3. Experimental measurements of the receiver prototype realized in 65-nm CMOS are then presented, followed by a brief comparison with other FM-UWB receiver realizations in Section 4.4. A comparison with other low-power receivers, and areas identified for future work are summarized in Section 4.5.

4.2 Proposed FM-UWB Receiver

An FM signal can only be detected indirectly (i.e., by converting it to a phase (PM) or amplitude modulated signal (AM) before detection) because physical systems are unable to read the instantaneous frequency of a carrier [2]. The conventional FM-UWB demodulator described in Section 2.6 first transforms FM to PM via a delay line with constant time delay, and then multiplies it with the original FM signal to yield an AM output.

In the proposed receiver, the FM-AM transformation is realized via an RF bandpass filter. Selecting a sub-band does not (ideally) affect the SNR of the received signal compared to processing the entire FM-UWB transmit signal. In fact, measurement of the prototype reveals that the received SNR and sensitivity for the sub-band FM-UWB receiver are on par with a conventional wideband receiver. However, the preamplifier in the receiver is made more power efficient by processing a sub-band rather than the entire band. In contrast to conventional UWB receivers which receive the entire transmitted signal (including noise and unwanted interference), the receiver proposed in this work [3] (see Figure 4.1) is designed according to the cognitive radio paradigm [4], in that it selects and operates on a sub-band of the FM-UWB RF signal. Potential interferers may be avoided in this way, and the received signal-to-noise ratio (SNR) is optimized to improve robustness by sub-band selection. The proposed receiver front-end consists of an RF preamplifier and pulse-shaping RF filter followed by an envelope detector. An IF amplifier and output buffer are also included on-chip to facilitate characterization of the receiver front-end.

The FM demodulation process used in the proposed receiver is illustrated in Figure 4.2. The carrier is initially moving down in frequency between time t_1 to t_2 (Figure 4.2(a) and Figure 4.2(b)). The bandpass response of the RF preamplifier, which is tuned to $f_C - \Delta f$, detects and amplifies the carrier energy within its passband. Otherwise, the carrier signal is attenuated. The narrowband response of the RF front-end therefore shapes the envelope of

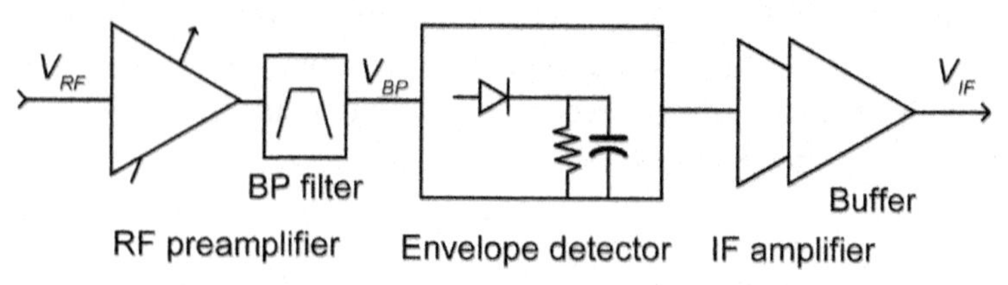

Figure 4.1 Proposed FM-UWB receiver front-end.

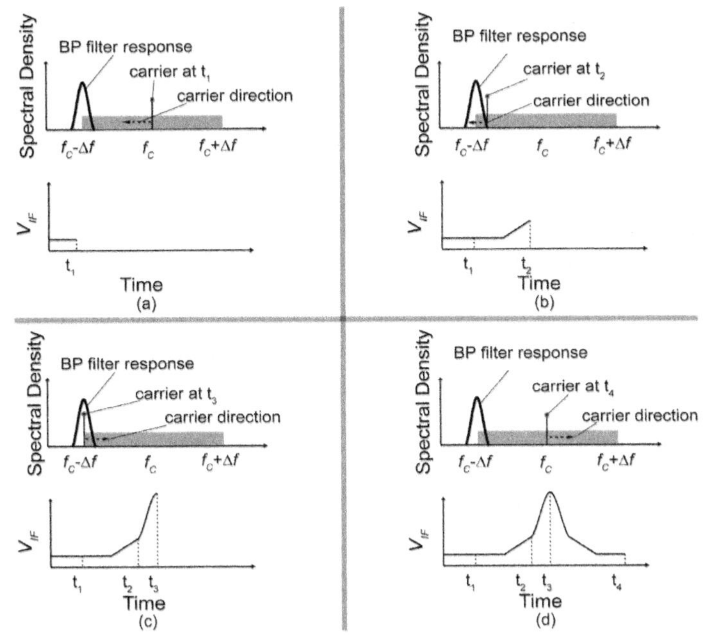

Figure 4.2 FM-UWB demodulation in sequence: (a) t_1, (b) t_2, (c) t_3 and (d) t_4, with the bandpass (BP) filter response tuned to low side of the FM-UWB signal.

the received FM into an AM signal according to the bandpass response of the preamplifier. Amplification is highest when the carrier is centered in the passband (i.e., V_{IF} at t_3 in Figure 4(c)), resulting in a peak in the amplitude of the demodulated signal. When the carrier is moving up in frequency (at t_4), the voltage at the IF output (V_{IF}) returns back to its original level. The carrier sweeps back and forth across the band at a rate determined by the (modulating) sub-carrier signal, resulting in a pulse train output at the IF. Note that if the passband is set to center frequency f_C, the output pulse rate is at double the data rate. A divide-by-2 circuit is therefore required in the baseband processing circuits.

Tuning the bandpass filter center frequency adds complexity to the receiver. However, it also permits optimization of the SNR when interference and noise are present. An automated tuning scheme has not been implemented, although calibration circuits that optimize the SNR and bit-error performance of the receiver could be devised and implemented on-chip. For example, in [5] an RF filter is tuned using a replica phase-locked loop, but has a disadvantage of a possible mismatch between the filter and its replica. In [6–8], the filter is initially configured as an oscillator so that the center frequency can be detected and tuned using a digital control loop. After initial calibration of the center frequency, the gain is reduced to suppress the oscillation. In the updated version of the receiver described in Chapter 5, this calibration circuit is included.

The demodulated signal (V_{IF}) is a periodic train of amplitude-modulated pulses. The shape of each pulse can be approximated as a Gaussian pulse g(t),

$$g(t) = e^{-(t^2/2b^2)}, \tag{4.1}$$

where the pulse width $(2b)$ is proportional to the bandwidth of the bandpass filter. The periodic train of pulses can be obtained by convolving (4.1) with a periodic train of impulses [9]

$$p(t) = \sum_{n=-\infty}^{\infty} \delta(t - nT), \tag{4.2}$$

where T is the pulse repetition rate. This resulting periodic signal in time is given by

$$h(t) = \sum_{n=-\infty}^{\infty} e^{-(t-nT)^2/2b^2}. \tag{4.3}$$

The Fourier transform of (4.3) yields the frequency spectrum of the filtered signal, which is:

$$H(f) = be^{-b^2 f^2/2} \sum_{m=-\infty}^{\infty} \delta\left(f - \frac{m}{T}\right). \tag{4.4}$$

Equation (4.4) predicts that the demodulated signal spectrum is also Gaussian, with a bandwidth inversely proportional to the width of the pulse (i.e., $2b$). It also consists of many equally spaced spectral lines (1/T apart). Filtering removes higher harmonics while preserving the fundamental frequency of the demodulated signal. Choosing the sub-carrier frequency so that the harmonics do not interfere with each other is needed when the FM-UWB system employs more than one sub-carrier frequency (e.g., in a multi-user system).

Results from simulation of the proposed receiver chain are shown in Figure 4.3. The FM-UWB signal has 500 MHz bandwidth and the sub-carrier frequency is 1 MHz ($f_C = 4.5$ GHz and -80 dBm RF power). The top part of the figure shows the carrier frequency versus time. The preamplifier and

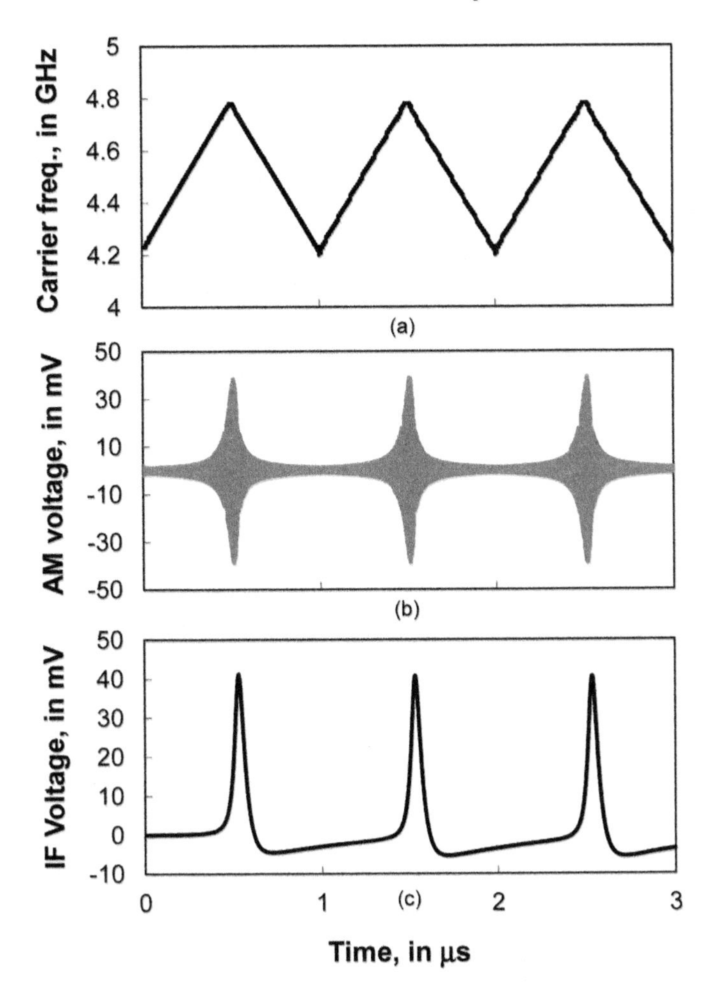

Figure 4.3 Signal transformation in the receiver chain: (a) input V_{RF} frequency, (b) AM signal at the preamplifier output, V_{BP}, and (c) detected output at IF, V_{IF}.

narrowband filter convert FM into AM as shown in Figure 4.3(b), and the envelope detector produces the pulse IF wave as shown in Figure 4.3(c). The simulated power spectra of the received signal before and after pulse shaping by the RF preamplifier are shown in Figure 4.4. The filter center frequency is tuned at the edge of the FM-UWB band, hence frequency near the edge see greater amplification than the rest of in-band signal. The output signal is amplitude modulated (AM) in the time domain, as shown in Figure 4.3(b).

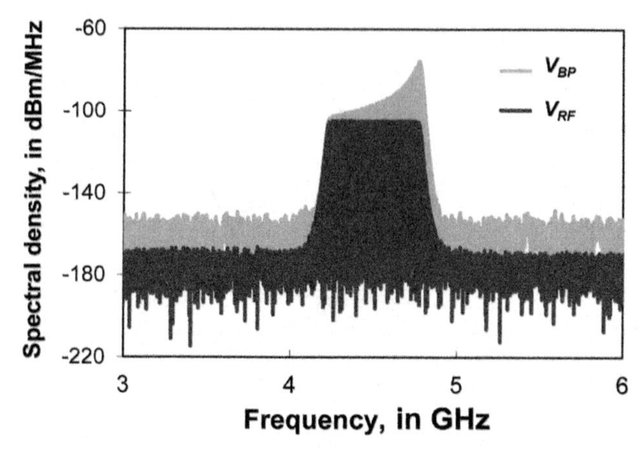

Figure 4.4 Simulated FM-UWB spectrum before (V_{RF}) and after the preamplifier (V_{BP}).

4.3 Circuit Designs

The prototype 4-5 GHz band FM-UWB receiver is designed and implemented in a 65-nm bulk CMOS [10]. The back-end interconnect scheme in this technology includes: 4 thin copper (Cu) metal layers, 2 medium-thick Cu layers, and 2 thick, Cu top-metal layers.

4.3.1 Regenerative Preamplifier and Filter

The input preamplifier should realize more than 30-dB RF gain at the lowest possible DC power consumption. It also requires good selectivity (i.e., 30-dB attenuation at 500-MHz offset) to select a 50-MHz sub-band. Positive feedback is therefore applied to the preamplifier to increase the voltage gain (but at the expense of lower bandwidth), thereby increasing its selectivity and sensitivity, as previously described in Section 2.7.2.

The regenerative preamplifier is shown in Figure 4.5. M_1 and M_2 form a transconductance amplifier, which drives the LC tank formed by L_D, C_{VAR} and parasitic capacitances at the drains of M_3, M_4, M_5 and M_6. Inductor L_D is 2 nH, with a self-resonant frequency (f_{RES}) of 15 GHz and peak-Q of 14.2 at 5.9 GHz. Varactor C_{VAR} is an NMOS accumulation mode varactor, which ranges from 1.1 pF to 0.3 pF as V_{CAP} is changed from 0.4 V to 1.5 V. V_{CAP} tunes the center frequency of the load resonator with across 3.8 to 5.1 GHz. Cascode transistors M_3, M_4, M_7 and M_8 increase isolation between the RF input and output, while cross-coupled transistors M_5 and M_6 form a positive feedback loop at the preamplifier output. The gate of cascode transistors,

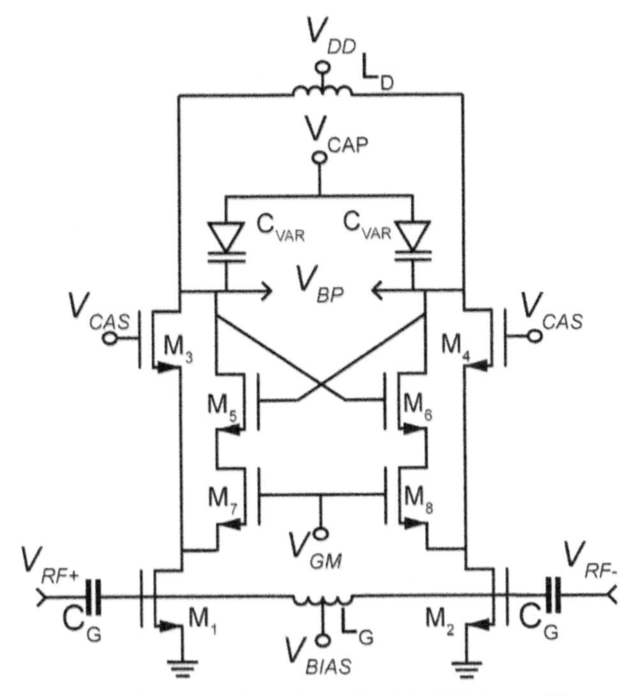

Figure 4.5 Simplified schematic of the RF preamplifier.

V_{CAS} is biased at 1 V, while V_{GM} is varied to control the current flowing through M_5 and M_6, thereby controlling the amount of positive feedback. The simulation results shown in Figure 4.6(a) illustrate the gain and bandwidth changes as V_{GM} is varied ($V_{GM1} > V_{GM2} > V_{GM3}$). Figure 4.6(b) shows the simulated change in the center frequency of the preamplifier's response as V_{CAP} is varied by ± 0.05 V ($V_{CAP1} < V_{CAP2} < V_{CAP3}$). The measured tuning range of the prototype is 3.8 GHz to 5.2 GHz. The preamplifier input is matched to 50 Ohm in the 4–5 GHz band by a conventional L-type network which consists of a 6.6-nH shunt input inductor (L_G), with a self-resonant frequency of 9.3 GHz and peak-Q of 11.2 at 5 GHz, and a 0.3-pF capacitor in series (vertical plate backend capacitor, C_G). L_G also provides a DC bias path (V_{BIAS}) and shunts any electrostatic discharge (ESD) events away from the transistor inputs. Inductor L_D and L_G are fully-symmetric layouts that provide the highest Q-factor in the smallest possible chip area [14].

Positive feedback around M_5 and M_6 enhances the selectivity of the amplifier, as seen from Figure 4.6(a). Positive feedback increases the output impedance (Z_O) of the preamplifier. Hence, the voltage gain, which is

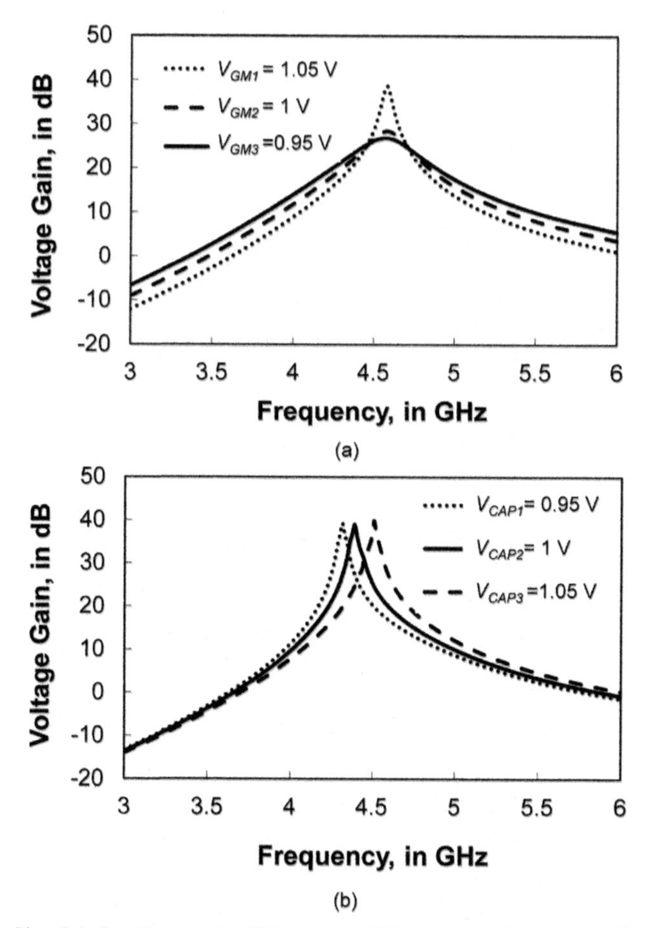

Figure 4.6 Simulated voltage gain of the preamplifier versus (a) *gm* control voltage V_{GM}, and (b) frequency control voltage V_{CAP}.

proportional to the product of transconductance (*gm*) and load impedance (Z_O) increases from 16 dB to 35 dB at the expense of bandwidth (520 MHz to 45 MHz) as the feedback increases. The preamplifier linearity in terms of input -1 dB compression point also drops from -20 dBm to -41 dBm, mostly due to the gain changes. As the feedback increases, the 50 Ω noise figure also increases from 4.6 dB (the main noise contribution comes from L_G, M_1 and M_2) to 7.3 dB (the main noise contribution comes from positive feedback branch M_3, M_4, M_5 and M_6) at an RF input frequency of 4.5 GHz. The preamplifier consumes 1.6 mW from the 1 V supply, which is 55% less power than the conventional FM-UWB preamplifier described in [15] for a comparable voltage gain.

4.3.2 Envelope Detector and IF Amplifier

Envelope detector (M_{N1} and M_{P1} in Figure 4.7) removes the carrier from the filtered RF signal. Source follower M_{N1} acts as a diode detector because the RF voltage swing on its source terminal is constrained by a 2-pF MOS capacitor, C_{ENV}. It is well-known that the envelope detector gain is less than unity for signal amplitudes below a certain threshold [16]. The simulated overall detector voltage gain as a function of input amplitude is shown in Figure 4.8. Signals smaller than the detector's input threshold of 250 mV

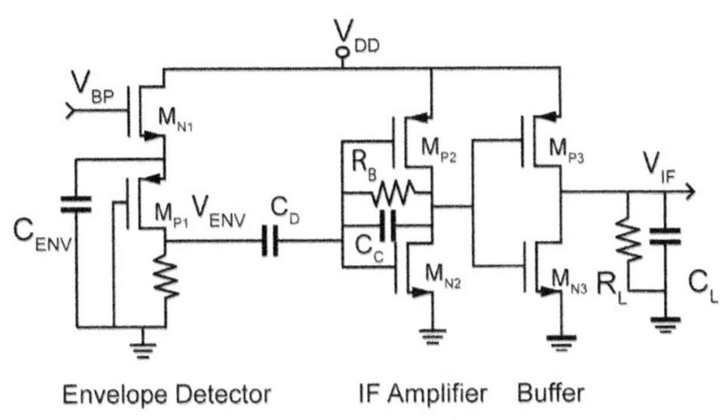

Figure 4.7 Schematic of the envelope detector (M_{N1}, M_{P1}), IF amplifier (M_{N2}, M_{P2}) and test buffer (M_{N3}, M_{P3}).

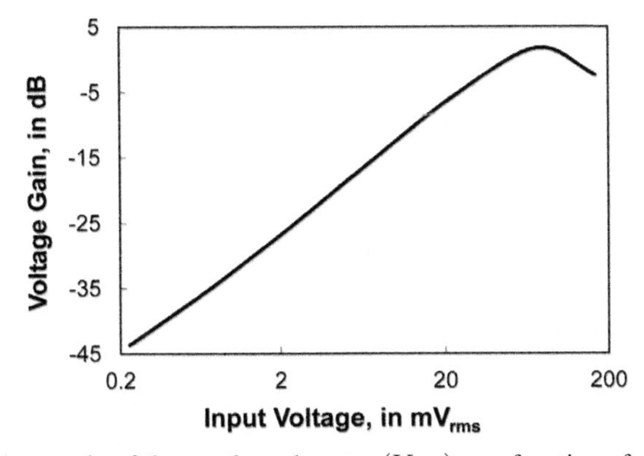

Figure 4.8 Voltage gain of the envelope detector (V_{ENV}) as a function of rms input voltage (V_{BP}).

are squared by the nonlinearity of the MOSFET. The gain peaks at around 80 mV_{rms}, and then drops at higher input amplitudes as M_{P1} drops out of saturation and into the triode operating region.

Common-gate transistor M_{P1} amplifies and buffers the detected signal to the IF amplifier input. The PMOS transistor reuses the bias current flowing through the envelope detector, thereby saving power. M_{P1} provides a voltage gain of 6 dB from source to drain and for extra filtering. The envelope detector is biased at 200 μA and is sensitive to variations in process, voltage and temperature (PVT), because its bias current depends on the threshold voltages of M_{N1} and M_{P1}. The sensitivity can be mitigated by adding a compensation circuit [17].

The envelope detector is followed by an intermediate frequency (IF) amplifier and output buffer stages, as shown in Figure 4.7. The detected envelope is amplified 20 dB by the CMOS IF amplifier M_{N2} and M_{P2}, which has a 100 kHz to 10 MHz bandwidth. Feedback resistor R_B provides a DC path for biasing the amplifier, while capacitor C_D provides bias isolation from the envelope detector. The output buffer (M_{N3} and M_{P3}) is capable of driving the 50-Ω input test instrumentation used for chip characterization. Transistors M_{N3} and M_{P3} are scaled six times wider than M_{N2} and M_{P2}, and symmetry in their physical layout is implemented to promote matching of the device parameters. The output buffer consumes 2.5 mW from a 1-V supply.

4.4 Measurement Results

The 0.3-mm^2 receiver prototype (see Figure 4.9) was realized in a production 65-nm CMOS technology. The 0.09-mm^2 input matching inductor (L_G) dominates the chip area. While the matching inductor is an essential part of the receiver input matching network (together with C_G), it could also be implemented off-chip, reducing the die area by 30%. The test die was characterized from on-wafer measurements and also mounted and wirebonded onto a printed circuit board (PCB) for characterization as shown in Figure 4.10. A high-frequency laminate PCB ($\tan\delta = 0.004$ at 5 GHz) is used to implement 0.16-dB/inch insertion loss 50-Ω transmission lines between the IC and coaxial test connectors located at the PCB periphery. The RF input bondwire length is 0.5 mm, resulting in a series inductance of approximately 0.5 nH and parasitic capacitance of 10 fF. The supply and bias lines are filtered using a 10.33-nF off-chip (i.e., on the PCB) decoupling capacitor in parallel with 15 pF of on-chip capacitance.

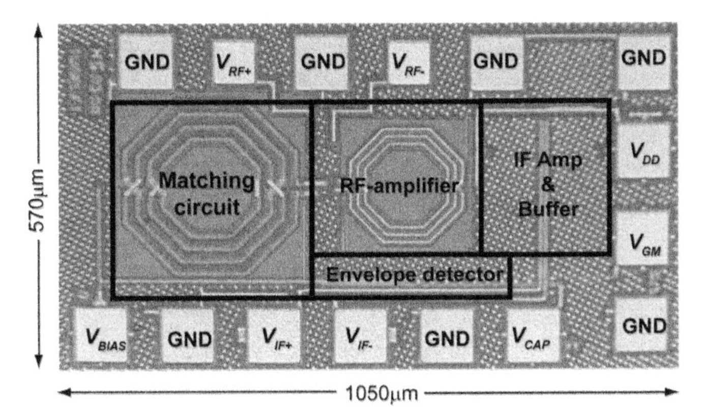

Figure 4.9 Micrograph of the 65-nm CMOS receiver prototype.

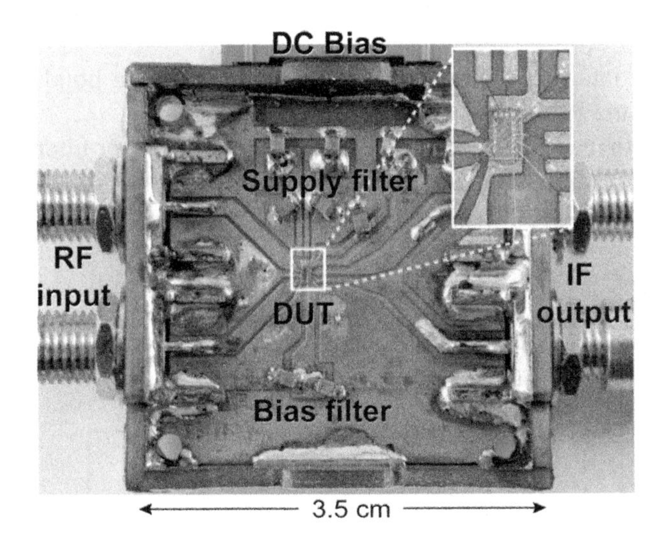

Figure 4.10 Assembled test PCB.

The RF input reflection coefficient (S_{11}) was characterized using a 4-port network analyzer [18] and on-wafer probing of the test circuit. The measured receiver input S_{11} from 4 to 5 GHz is plotted in Figure 4.11 with the positive feedback turned off. In this condition, the circuit behaves like a conventional cascode amplifier. The S_{11} measured with on-wafer RF probes (SOLT calibrated to the probe tips) and for the die mounted on the PCB shows a good agreement with the post-layout simulation results. The slight shift in minimum S_{11} of 150 MHz observed for the packaged device (SOLT calibrated

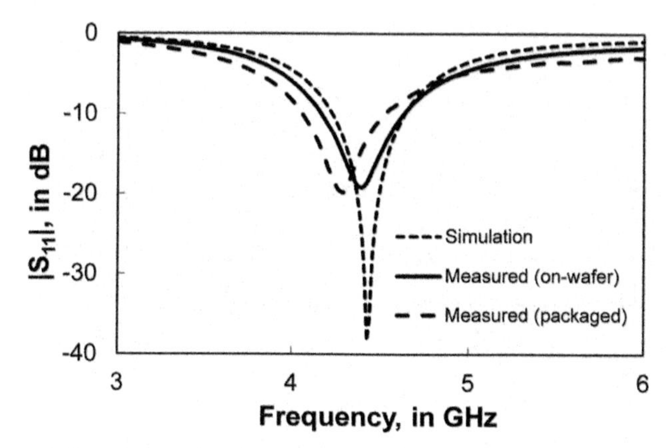

Figure 4.11 Measured and simulated receiver S_{11} (no positive feedback).

to the connector interface) is due to parasitics added by the bondwire and RF-connector to the RF input.

It is difficult to measure the center frequency of the receiver filter directly at the preamplifier output without an extra RF buffer, because of its high output impedance (>1 kΩ). When the positive feedback is active, the output signal leaks back to the source of the cascode transistor and then to the RF input terminal through C_{GD} of the input transistor. The null normally seen in S_{11} without feedback is therefore disturbed, and can be controlled by changing V_{CAP} as shown in Figure 4.12. Compared to Figure 4.11 (no feedback), the

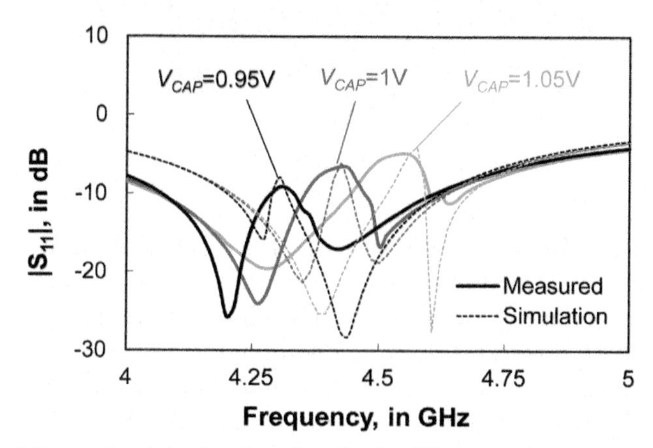

Figure 4.12 Measured and simulated receiver S_{11} for different settings of V_{CAP} when positive feedback is applied.

positive feedback degrades the input match in the vicinity of the bandpass center frequency. The effect of tuning V_{CAP} on the measured S_{11} can be seen (tuning sensitivity). Figure 4.13 shows the measured center frequency of the receiver as a function of V_{CAP}. The measured frequency range is 3.8–5.2 GHz with a curve that corresponds to the shape of the varactor C-V function.

In testing, the receiver was tuned to its optimum sensitivity by adjusting V_{GM} and V_{CAP} to change the preamplifier's tank Q and its sub-band frequency, respectively. Firstly, the amplifier gain was increased via V_{GM} until oscillation commences, which causes an abrupt change in the V_{IF} output DC bias. Then, V_{GM} was slowly decreased until oscillation stops, which can be observed easily as the DC output drops back to its original value. The receiver is now operating near its highest voltage gain and selectivity.

The test setup used to measure the receiver sensitivity is shown in Figure 4.14. The FM-UWB transmitter is driven by a random binary data

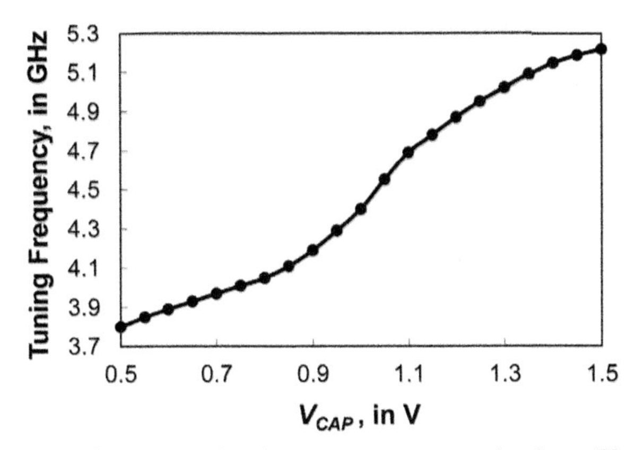

Figure 4.13 Measured tuning range versus control voltage, V_{CAP}.

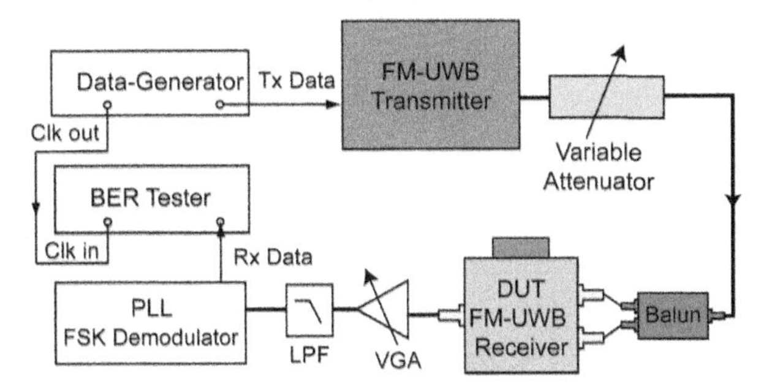

Figure 4.14 Block diagram of measurement setup.

source and produces a wideband FM signal at −14 dBm output power. The received signal level is adjusted using a variable attenuator to emulate the propagation losses of an RF link. A balun converts the single-ended signal to differential at the FM-UWB receiver chip RF input. A variable-gain amplifier (VGA) amplifies the IF output signal to 0.5 $V_{p\text{-}p}$. The FSK IF signal is then low-pass filtered, and demodulated using a 74HCT9046 phase-locked loop, and then fed to the error detector to measure the bit-error rate (BER).

The measured spectra at the FM-UWB transmitter output are plotted in Figure 4.15 for peak power spectral densities of −40, −55 and −70 dBm/MHz. The transmitted signal occupies 500-MHz bandwidth. Figure 4.16 shows the received sub-carrier output waveform at 1 MHz for a measured RF input power of −50 dBm. The pulse train seen at the IF output agrees well with the analysis presented by Section 4.2 and the simulation results shown in Figure 4.3(c).

The measured bit-error rate of the receiver as a function of the received input power is plotted in Figure 4.17. The receiver is tuned to a center frequency of 4.3 GHz and the transmitted signal band ranges from 4.25–4.75 GHz. The receiver sensitivities (10^{-3} BER) at data rates of 50, 100 and 200 kbit/s are −87, −84 and −78 dBm, respectively. Deviation from the theoretical BER 'waterfall' shape for the measured BER curves is likely caused by the FSK (PLL) demodulator (not on-chip), which displays a threshold effect before reliably demodulating the received signal. The measured results imply a 4 dB link margin for the receiver running at 100 kbit/s (some of which will be taken

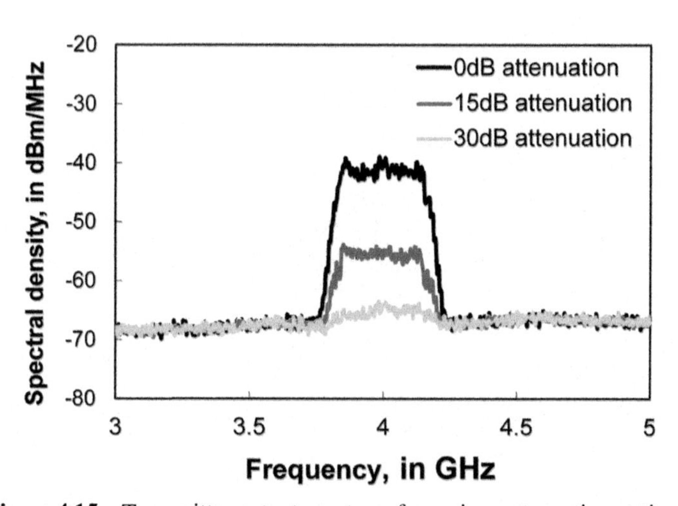

Figure 4.15 Transmitter output spectrum for various attenuation settings.

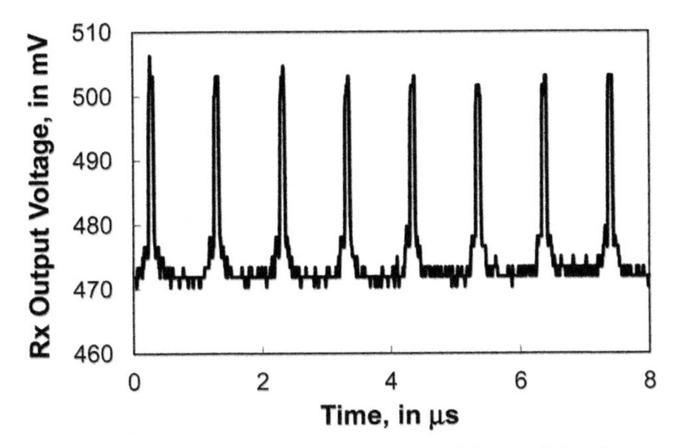

Figure 4.16 Receiver output pulses measured for 1 MHz sub-carrier.

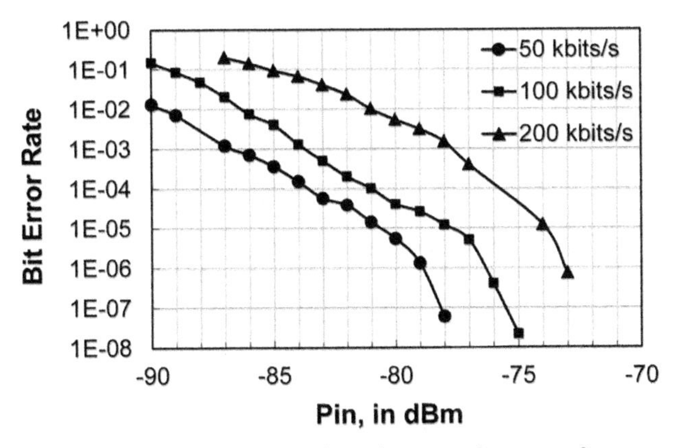

Figure 4.17 Measured bit-error rate at various data rates for center frequency of 4.3 GHz.

up by the baseband processor), given that −80 dBm sensitivity is desired for the 10-m range.

Figure 4.18 shows the measured BER when the receiver center frequency is tuned to 3.8, 4.3 and 4.8 GHz. The sensitivity of the receiver (10^{-3} BER) within the tuning range varies from −82 to −86 dBm at a data rate of 100 kbit/s. The variation of the sensitivity is contributed by output noise variation of the receiver at different center frequencies, and also variation of the input matching condition. Figure 4.19 illustrates the measured BER for single-tone interference applied out of band at 2.4 GHz and 5.8 GHz, and in-band at 4.5 GHz. The received input power is −80 dBm at a data rate of 100 kbit/s. The FM-UWB receiver is able to tolerate in-band interference that is 30-dB

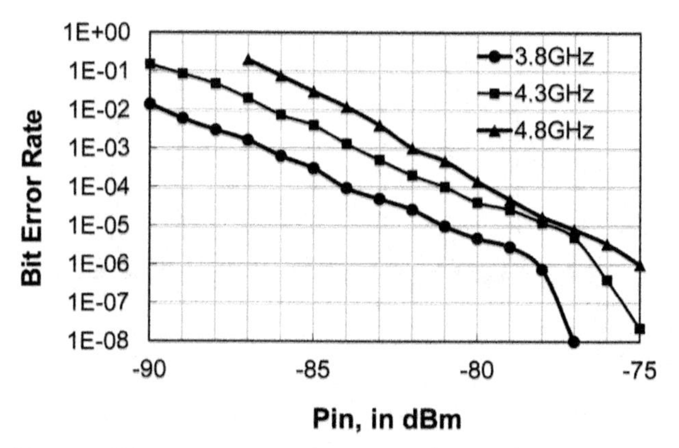

Figure 4.18 Measured bit-error rate at various center frequencies for data rate of 100 kbit/s.

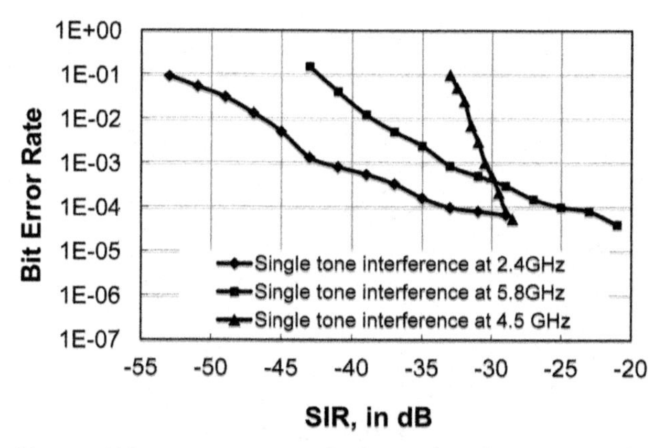

Figure 4.19 Measured bit-error rate versus single-tone interference strength for interference at 2.4, 4.5 and 5.8 GHz at data rate of 100 kbit/s and center frequency of 4.3 GHz.

stronger than the input signal power. The out-of-band interference rejection can be improved by adding a preselect filter ahead of the receiver RF input.

Frequency selective fading due to multipath is simulated by splitting the transmit signal into two paths which have different lengths ($\Delta l = 1.2$ m). The two paths are combined again at the receiver, yielding the spectrum shown in Figure 4.20. The peaks and nulls emulate the nulls and peaks in the received signal that might be encountered due to propagation path delay differences (i.e., multipath effect). The receiver center frequency is tuned across 4 GHz to 5 GHz and the receiver sensitivity measured in the transmission 'windows' of the emulated channel. The experiment demonstrates the receiver agility when

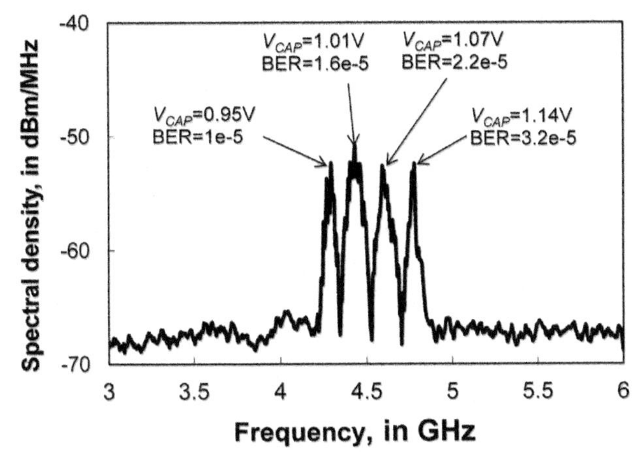

Figure 4.20 Transmitted output spectrum when frequency selective fading occurs due to multipath.

frequency selective fading is present in the channel. However, a mechanism for active tuning is required in practice, which is a topic for future study and development. Adaptive control circuitry could be added to the receiver to initiate a frequency hopping scheme to scan the receiver across the UWB for a suitable transmission window when the BER is high due to multipath or strong interference.

Consuming just 2.2 mW at 100 kbit/s, the energy efficiency of the receiver front-end is 22 nJ/bit, which is an improvement of almost an order of magnitude compared to earlier FM-UWB designs. Overall, the receiver is more power efficient by processing a sub-band, rather than the entire UWB band. The receiver performance is summarized and also compared with other FM-UWB receiver in Table 4.1.

Table 4.1 FM-UWB receiver performance comparison

Parameters	This Work	[19]	[20]
Technology	CMOS 65 nm	SiGe BiCMOS 80 nm	SiGe BiCMOS 250 nm
RF band, in GHz	4–5	3.1–4.9	7.2–7.7
Power consumption, in mW	2.2	10	9.1
Receiver sensitivity, in dBm	−84	−46	−86.8
Data rate, in kbit/s	100	100	50
Energy efficiency, in nJ/bit	22	100	180
Active area, in mm^2	0.3	0.72	0.88

4.5 Conclusion and Summary

A new architecture for detecting FM-UWB signals was described and validated in this chapter. Positive feedback in a controlled way was employed to realize good selectivity, high gain, and FM-to-AM transformation in a single circuit block, which improves receiver sensitivity and lowers power consumption. The ability to process a sub-band of the received FM-UWB signal could also be used to avoid interference and optimize the received SNR. The 0.3-mm^2 (active area) receiver front-end prototype realized in a production 65-nm CMOS technology demonstrated −84 dBm input sensitivity at 100 kbit/s, while consuming 2.2 mW from a 1-V supply for an energy efficiency of 22 nJ/bit. The energy efficiency of this receiver prototype is still slightly higher than the desired specification. An improved prototype described in next chapter addresses this shortcoming.

The receiver is compared with other low-power, low-data-rate receivers in Table 4.2. The receiver designed in this work consumes the least power and chip area among the FM-UWB receivers, while realizing comparable sensitivity. The wake-up receiver [22] consumes less power than an FM-UWB design, however, the sensitivity is 12 dB poorer, at −72 dBm. The wake-up receiver also requires a high-selectivity, passive RF preselect filter (e.g., SAW or BAW high-Q bandpass) to attenuate potential interferers to a tolerable level. The super-regenerative receiver in [23] achieves high energy efficiency and sensitivity by using BFSK modulation. However, the narrowband receiver is susceptible to multipath fading or interference effects because the front-end operates at a fixed input frequency and is not tunable. The UWB receiver in [24] also realizes good energy efficiency by receiver duty cycling, but efficiency

Table 4.2 Low-power receiver performance comparison

Parameters	This Work	[21]	[22]	[23]	[24]	[25]
Standard	FM-UWB	Zigbee	Wake-up	Super-Regen.	UWB	UWB
Technology	CMOS 65 nm	CMOS 180 nm	CMOS 90 nm	CMOS 180 nm	CMOS 90 nm	CMOS 130 nm
Modulation	2-FSK	QPSK	OOK	BFSK	PPM	BPSK
RF band, in GHz	4–5	2.4	2	2.4	4.4	0–0.96
Power Consumption, in mW	2.2	26.5	0.052	0.215	35.8	3.3
Receiver sensitivity, in dBm	−84	−101	−72	−86	−99	−55
Data rate, in kbit/s	100	250	100	250	100	1300
Energy efficiency, in nJ/bit	22	106	0.52	0.84	2.5	3.3
Active area, in mm^2	0.3	3.8	0.1	0.55	1	4.52

suffers when continuous data streaming is required. The other UWB receiver in [25] operates at 1.3 Mbit/s, but has poor sensitivity (−55 dBm) and is confined to RF inputs below approximately 1 GHz. In summary, the FM-UWB receiver designed in this work offers a compact implementation with better coexistence and robustness than other low-power receiver technologies in continuous data streaming applications at rates on the order of 100 kbit/s.

The results of this study validate the regenerative FM-UWB receiver front-end architecture. Automatic digital tuning is required to control the positive feedback loop in a practical receiver. Simultaneous frequency and Q-tuning for the RF preamplifier [5–8] can also be applied to this receiver in the same way it has been applied to monolithic filters. A fully-digital front-end calibration scheme and digital baseband processing are preferred because they are more amenable to scaling in future CMOS process generations [26, 27]. A fully-integrated receiver incorporating baseband processing to demodulate the sub-carrier will therefore be implemented to further validate the receiver concept and benchmark its overall performance.

The following chapter will describe a full FM-UWB transceiver (i.e., transmitter, receiver, baseband modulator, and demodulator) prototype realized in CMOS. The design has lower power consumption compared to the designs described in Chapter 3 and this chapter. Both the receiver and transmitter are integrated on-chip, along with other peripheral circuitry such as biasing, control logic and calibration.

References

[1] D. C. Daly, A. P. Chandrakasan, "An energy efficient OOK transceiver for wireless sensor networks," *IEEE Journal of Solid State Circuits*, Vol. 42, No. 5, pp. 1003–1011, May 2007.

[2] M. H. L. Kouwenhoven, *High-performance frequency demodulation systems*. Delft University Press: Delft, the Netherlands, 1998.

[3] N. Saputra, J. R. Long, J. J. Pekarik, "A 2.2 mW regenerative FM-UWB receiver in 65 nm CMOS," *proc. IEEE RFIC Symposium*, May 2010, pp. 193–196.

[4] A. Sahai, R. Tandra, S. M. Mishra, and N. Hoven, "Fundamental design tradeoffs in cognitive radio systems," *proceeding of TAPAS*, Vol. 222, No. 2. Aug. 2006.

[5] D. Li, and Y. P. Tsividis, "A loss-control feedback loop for VCO indirect tuning of RF integrated filters," *IEEE Transaction on Circuits*

and Systems-II: Analog and Digital Signal Processing, Vol. 47, No. 3, pp. 169–175, March 2000.

[6] J. K. Nakaska and J. W. Haslett, "2 GHz automatically tuned Q-enhanced CMOS bandpass filter, "*proc. of IEEE MTT-S International Microwave Symposium*, June 2007, pp. 1599–1602.

[7] H. Liu, and A. I. Karsilayan, "An accurate automatic tuning scheme for high-Q continuous-time bandpass filters based on amplitude comparison," *IEEE TCAS-II: Analog and Digital Signal Processing*, Vol. 50, No. 8, pp. 415–423, Aug. 2008.

[8] C. DeVries and R. Mason, "A 0.18um CMOS, high Q-enhanced bandpass filter with direct digital tuning," *proc. of IEEE Custom Integrated Circuits Conference*, May 2002, pp. 279–282.

[9] B. Porat, *A course in digital signal processing*, New York, NY: John Wiley & Sons, 1997.

[10] Z. Luo, A. Steegen, M. Eller, et al., "High performance and low power transistors integrated in 65 nm bulk CMOS technology," *IEDM Technical Digest*, pp. 661–664, Dec. 2004.

[11] D. M. Binkley, *Tradeoffs and optimization in analog CMOS design*, West Sussex, England: John Wiley & Sons, 2008.

[12] E. H. Armstrong, "Some recent developments in the audion receiver," *Proceedings of the IRE* Vol. 3, No. 9, pp. 215–247, Sept. 1915.

[13] Armstrong, E. H., U.S. Patent 1,113,149, Wireless receiving system, 1914.

[14] M. Danesh and J. R. Long, "A differentially-driven symmetric microstrip inductor," *IEEE Transactions in Microwaves Theory and Techniques*, Vol. 50, No. 1, pp. 332–341, Jan. 2002.

[15] Y. Zhao, G. van Veenendaal, H. Bonakdar, J. F. M. Gerrits, and J. R. Long, "3.6mW, 30dB gain preamplifiers for an FM-UWB receiver," *Proceeding of IEEE BCTM*, Oct. 2008, pp. 216–219.

[16] R. G. Meyer, "Low-power monolithic RF peak detector analysis," *IEEE Journal of Solid State Circuits*, Vol. 30, No.1, pp. 65–67, Jan. 1995.

[17] Y. Tsugita, K. Ueno, T. Hirose, T. Asai, and Y. Amemiya, "On-chip PVT compensation techniques for low-voltage CMOS digital LSIs," *Proc. International Symposium on Circuits and Systems*, May, 2009, pp. 1565–1568.

[18] J. Dunsmore, "New measurement results and models for non-linear differential amplifier characterization," *Proc. European Microwave Conference*, Vol. 2, Oct. 2004, pp. 689–692.

[19] J. F. M Gerrits, J. R. Farserotu, and J. R. Long, “A wideband FM demodulator for a low-complexity FM-UWB receiver,” *Proceeding of the 9th European Conference on Wireless Technology*, Sept. 2006, pp. 99–102.

[20] Y. Zhao, Y. Dong, J. F. M. Gerrits, G. van Veenendaal, J. R.Long, and J. R. Farserotu, “A short range, low data rate, 7.2 GHz–7.7 GHz FM-UWB receiver front-End,” *IEEE Journal of Solid State Circuits*, Vol. 44, No. 7, pp. 1872–1881, July 2009.

[21] W. Kluge, F. Poegel, H. Roller, M. Lange, T. Ferchland, L. Dathe, and D. Eggert, “A fully integrated 2.4 GHz IEEE 802.14.4-compliant transceiver for Zigbee^{TM} applications,” *IEEE. Journal of Solid-State Circuits*, Vol. 41, No. 12, pp. 2767–2775, Dec. 2006.

[22] N. M. Pletcher, S. Gambini and J. M. Rabaey, “A 52 μW, wake-up receiver with −72 dBm sensitivity using uncertain-IF architecture,” *IEEE. Journal of Solid-State Circuits*, Vol. 44, No. 1, pp. 269–280, Jan. 2009.

[23] J. Ayers, K. Mayaram, T. S. Fiez, “An ultralow-power receiver for wireless sensor networks,” *IEEE Journal of Solid State Circuits*, Vol. 45, No. 9, pp. 1759–1769, Sept. 2010.

[24] F. S. Lee and A. Chandrakasan, “A 2.5 nJ/b 0.65 V 3-to-5 GHz subbanded UWB receiver in 90nm CMOS,” *IEEE Int. Solid-State Circuits Conf. Dig. Tech. Papers*, Feb. 2007, pp. 116–590.

[25] N. van Helleputte, M. Verhelst, W. Dehaene, G. Gielen, “A reconfigurable, 130 nm CMOS 108 pJ/pulse, fully Integrated IR-UWB receiver for communication and precise ranging,” *IEEE Journal of Solid State Circuits*, Vol. 45, No. 1, pp. 69–83, Jan. 2010.

[26] E. Grayver, and B. Daneshrad, “A low-power all-digital FSK receiver for space applications,” *IEEE Transaction on Communications*, Vol. 49, No. 5, pp. 911–921, May 2001.

[27] R. B. Staszewski, and P. T. Balsara, *All-digital frequency synthesizer in deep-submicron CMOS*, New Jersey: John Wiley & Sons, Inc., Sept. 2006.

5

FM-UWB Transceiver

5.1 Introduction

In Chapters 3 and 4, FM-UWB transmitter and receiver prototypes were described. The final goal is to integrate them together into a full FM-UWB transceiver. The main building blocks in the transmitter are sub-carrier and RF oscillators, and a power amplifier, while the receiver consists of a low-noise amplifier, an FM demodulator, and an FSK detector, as shown in Figure 5.1. The objective of this work is to reach an energy efficiency below 10 nJ/bit for each of the transmitter and receiver. Integrating transmitter and receiver together onto the same IC saves power, reduces the number of I/O pads, and the overall chip area, as they share building blocks such as biasing, clock generation and calibration circuitry.

Depending upon the application, the transmitter and receiver may be active at the same time, which require duplexing so that transmitter and receiver do not interfere with each other. A communication protocol is therefore required to manage wireless transmission between devices. Alternately, time duplexing may be used so that transmitting and receiving sections are not active at the same time. In a time-duplexed FM-UWB transceiver, the transmitter (or receiver) is turned 'off' during reception (or transmission) of data in order to save power. An external duplexer (for time switching) or diplexer (for transmitter and receiver frequency separation) is therefore required between the antenna and the transceiver IC. An RF filter (which may be incorporated in the antenna design) attenuates out of band harmonics and potential RF blockers.

The FM-UWB transceiver test chip architecture and building blocks are described in Section 5.2 of this chapter. Design details of each of the transceiver circuit blocks are discussed in Section 5.3. Experimental results of the transceiver prototype realized in 90 nm CMOS are discussed and compared

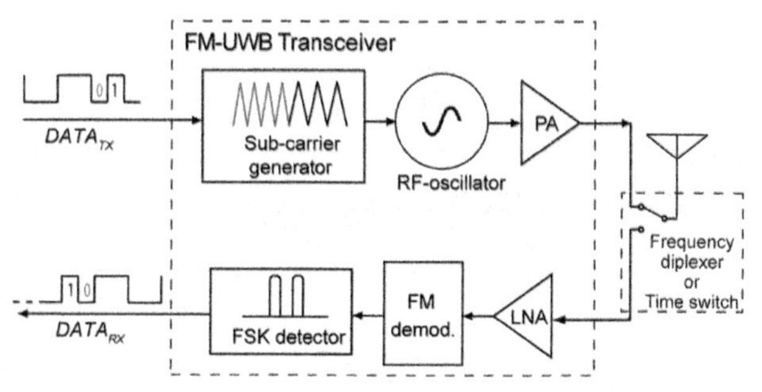

Figure 5.1 Block diagram of a typical FM-UWB transceiver.

with other FM-UWB transceivers in Section 5.4. Conclusions, a chapter summary, and areas identified for future work are summarized in Section 5.5.

5.2 Transceiver Architecture

The proposed fully-integrated and frequency-agile FM-UWB transceiver is shown in Figure 5.2. Generation of the signal in the FM-UWB transmitter is a two-step process, as described previously. Firstly, the triangular sub-carrier is generated using a relaxation oscillator, and FSK modulated by the binary transmit data. The triangular output voltage from the relaxation oscillator is converted to a current by transconductor G_M. A current-controlled ring oscillator (ICO) generates a 3-phase RF carrier operating at one-third of the

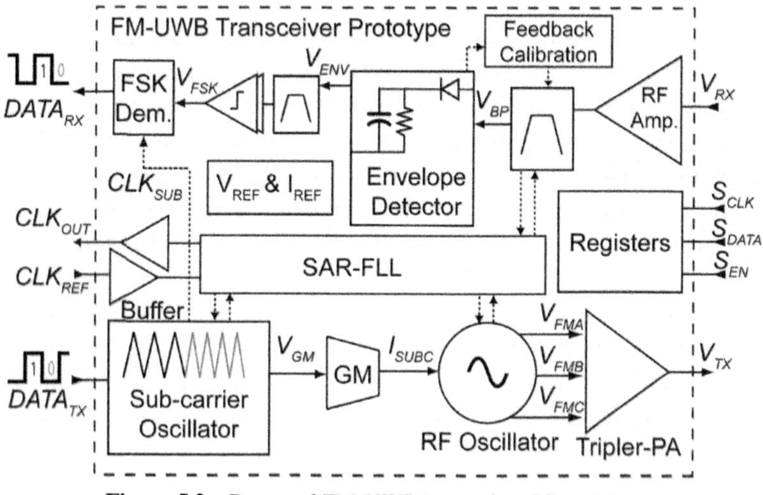

Figure 5.2 Proposed FM-UWB transceiver IC prototype.

output frequency (RF oscillator in Figure 5.2). The transmit signal is tripled in frequency and amplified for transmission by the tripler/power amplifier that drives the antenna via an on-chip output matching network.

On the receiver side, the received carrier signal from the antenna (V_{RX}) is amplified and filtered by a new regenerative RF amplifier design. An envelope detector recovers the FSK-modulated sub-carrier signal. The detected signal is bandpass filtered, amplified at IF, quantized by a 1-bit limiter, and then FSK demodulated to produce the received data bit stream. A current-reuse technique is employed extensively in the transceiver to conserve power. For example, supply current is shared between the frequency tripler and PA circuits in the transmitter, and also between the regenerative LNA and envelope detector in the receiver front-end.

An on-chip successive approximation register frequency-locked loop (SAR-FLL) calibrates the sub-carrier, RF oscillator, and regenerative amplifier center frequencies. Positive feedback is adjusted during calibration to tune the regenerative amplifier to reach its highest gain and sensitivity. Voltage and current references are also implemented on-chip to generate PVT insensitive references, and for general biasing. DACs are implemented in various building blocks to control the transceiver operating mode and bias. Serial registers are also included to supply the digital words that control calibration and tuning of the transceiver via a number of DACs (i.e., C-DACs and I-DACs) implemented in the various sub-blocks. A buffer is added to monitor the IF, various clocks, and digital data signals for test purposes. A mirror of the reference current (typical 600 nA) can also be measured to characterize its sensitivity to supply voltage and temperature variations.

5.3 Transceiver Circuit Design

The prototype FM-UWB transceiver is implemented in a 90-nm bulk-CMOS technology with thick metal and metal-insulator-metal (MIM) capacitor options for RF circuit applications [1]. The all-copper interconnect scheme (bottom to top of the stack) consists of 5 thin, two medium-thick, and one thick top metal. In this prototype, a single 1-V supply is chosen. Small modifications and improved circuit layouts are implemented across the blocks described in the previous chapter, such as the voltage/current reference, SAR-FLL, and the sub-carrier oscillator. The ICO is modified and optimized to operate at one-third the RF carrier frequency and has 3-phase outputs. The receiver blocks and the PA are completely new. The following section describes the new or upgraded building blocks in the transceiver that were not described in previously in Chapters 3 or 4.

5.3.1 RF Current-Controlled Ring Oscillator

As discussed in Section 3.2.1, a current-controlled ring oscillator (ICO) was chosen for the RF transmit oscillator because of its advantages such as a linear and wide tuning range, scalability with technology, scalable control current, and low susceptibility to noise present on the supply or ground lines. A three-stage ring oscillator is used, and it operates at one-third of the desired RF oscillation frequency to minimize power consumption. By operating at one-third of the carrier, unwanted pulling of the ICO frequency from a strong interferer at the transmitter output is mitigated. Furthermore, the required tuning range of the ICO is reduced to one-third, and the amplitude variation is much less than the earlier version of the ICO described in Section 3.2.1. These changes improve the frequency versus current linearity.

Each stage of the ICO consists of an NMOS driver with PMOS active load, as shown in Figure 5.3. The PMOS (M_{PR1}-M_{PR3}) source voltage (V_{REG}) is regulated to 0.75 V by opamp A_2, via a feedback loop that derives the desired

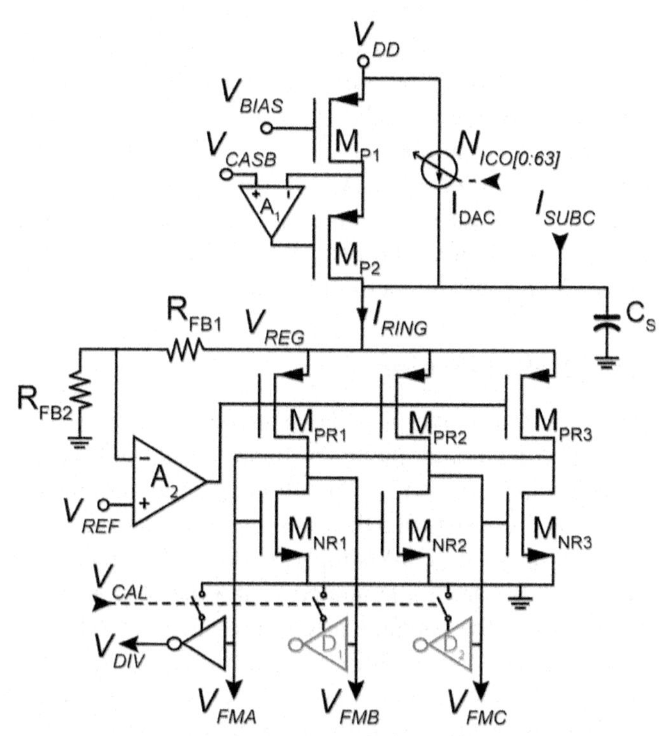

Figure 5.3 Current-controlled RF oscillator (RF-ICO) schematic.

supply voltage from a temperature-compensated voltage reference (typical $V_{REF} = 0.5\,\mathrm{V}$). Regulation of the ring oscillator supply desensitizes it to external variations in the supply voltage, V_{DD}. As the current that supplies the ICO (I_{RING}) increases (while maintaining V_{REG} constant), the ICO output frequency (f_{osc}) increases proportionally. Current I_{RING} is the sum of the bias current from M_{P1}, a 6-bit current DAC (I_{DAC}), and the sub-carrier modulation current I_{SUBC}, all of which are supplied by PMOS current sources. The output impedance of the current source is enhanced by the feedback amplifier A_1, which improves the accuracy of the current mirror, and yields 65-dB power supply rejection from DC up to 10 kHz.

The three output phases generated by the ICO (V_{FMA}, V_{FMB}, and V_{FMC}) are 120° apart. One of the phases (V_{FMA}) is buffered, frequency-divided by 1024, and calibrated using an FLL. The transistor sizes in each stage of the ICO are equal and drawn symmetrically in the layout to minimize mismatch between the output phases. The capacitive loading on each stage is also balanced, i.e., dummy buffers (D_1 and D_2) are implemented. A 1-pF decoupling capacitor (C_S) acts as a charge reservoir which reduces fluctuations in node voltage V_{REG}, and also filters out high-frequency supply noise. The current DAC is controlled by the 6-bit word N_{ICO}, and it is calibrated to provide a frequency tuning range of 1–1.7 GHz, and a controllable, modulated RF output bandwidth from 88–320 MHz. The ICO (incl. opamps and bias circuits) consumes 90 μA of DC current when it is oscillating at 1.33 GHz (one-third of the RF carrier at 4 GHz). The tuning sensitivity of the ICO is 38 MHz/μA.

5.3.2 Frequency Tripler and Power Amplifier

The output signal of the ICO is buffered into a full swing, rail-to-rail clock signal. Each phase is multiplied with the others by a series of NMOS switches, creating a times-three frequency multiplier as shown in Figure 5.4. Finally, a CMOS power amplifier (PA) amplifies the frequency-tripled signal, and drives a 50 Ω load via an LC matching network consisting of a 4-nH bondwire inductor and 0.1–0.3 pF shunt capacitance controlled by a DAC (C_{BD2}). Absorbing the wirebond in the output matching network reduces the number of components, and minimizes the chip area and cost. A large proportion (>40%) of the total transceiver power is dedicated to the PA stage to maximize the link span between transmitter and receiver. Power consumption is minimized by having the frequency multiplier and the PA share the same bias current (i.e.,

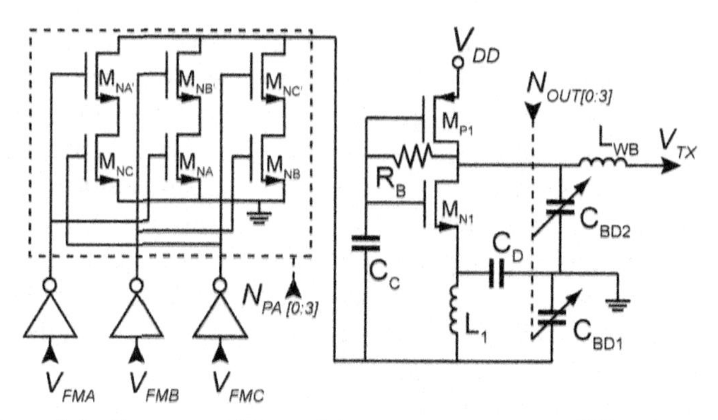

Figure 5.4 Frequency tripler and current reuse power amplifier schematic.

current reuse). The DC current consumed by the PA flows via L_1 to the tripler, with C_D providing the AC ground for M_{N1} of the PA.

The frequency tripler consists of 3 NMOS switching stages, where each series switch (e.g., $M_{NA'}$ and M_{NC}) realizes a logical AND function driven by 2 of the 3 CCO phases. Drain currents flow in the output of each switch only when pulses driving the gate inputs overlap, shrinking the duty cycle of the drain current by a factor of 3 (i.e., to 16.7%) compared to the voltages applied at each gate (50% duty cycle square waves). The three-phase drain currents from switch outputs $M_{NA'}$, $M_{NB'}$ and $M_{NC'}$ are summing via a wired-OR connection to produce a 50% duty cycle carrier at triple the input frequency. The tripler sizes are controlled by 2-bit digital word N_{PA} to adjust the overall gain and output power of the PA stage over a 5 dB range. The tripler current generates a voltage across resonant tank L_1 and C_{BD1}, which drives the CMOS PA (M_{N1} and M_{P1}) through coupling capacitor C_C.

The typical FM-UWB bandwidth is only 500 MHz, but the transmitter must cover the frequency range from 3–5 GHz. The LC matching network at the PA output has a -3 dB bandwidth of less than 1 GHz. The 4 overlapping transmit sub-bands are employed to cover the entire frequency range specified, and ensure efficient coupling of the transmitted power to the output as shown in Figure 5.5. The 2-bit control word N_{OUT} is employed to vary the capacitor DAC in the resonant circuit (L_1, C_{BD1}) and the output matching network (L_{WB}, C_{BD2}) to select the operating sub-band. The 10-pF decoupling capacitor (C_D) that AC grounds the CMOS PA is implemented using a parallel stack of MOS, MIM, and back-end metal-oxide-metal (MOM) capacitors to maximize the capacitance density and minimize the chip area. The tripler-PA consumes 490 μW (78% of the total) and produces up to -8.8 dBm (simulated

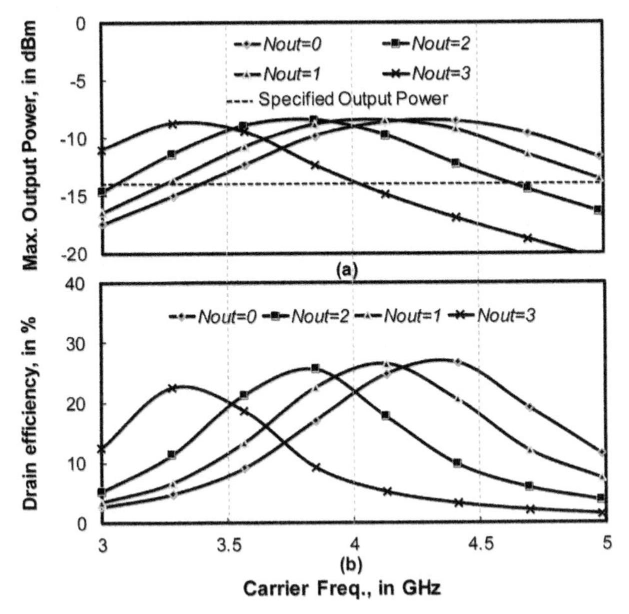

Figure 5.5 Simulated maximum output power (a) and drain efficiency (b) of the PA versus frequency for various setting of N_{OUT}.

maximum) output power. The simulated peak drain efficiency of the tripler-PA is 27%.

5.3.3 Transconductance Amplifier

The differential push-pull transconductance amplifier (GM block) shown in Figure 5.6 translates the sub-carrier voltage to a current that modulates the transmit RF-ICO. Amplifiers A_1 and A_2 form a push-pull buffer that drives on-chip polyresistor R_{GM}. The resulting current

$$I_{inj} = \frac{V_{GM} - V_{REFM}}{R_{GM}}, \quad (5.1)$$

is injected into the sources of transistors M_{S2} and M_{S4}. When the frequency modulation is turned on, V_{GM} is a triangular wave voltage generated by the sub-carrier oscillator (see Section 3.2.3), otherwise V_{GM} is equal to V_{REFM} so that no current is injected. Internal reference voltage V_{REFM} has a typical value of 0.4 V. The injected input current split between PMOS M_{P1} and NMOS M_{N1}, and mirrored to form a push-pull output current with no DC offset. The output current magnitude is set by a 6-bit word (N_{GM}) that controls the current

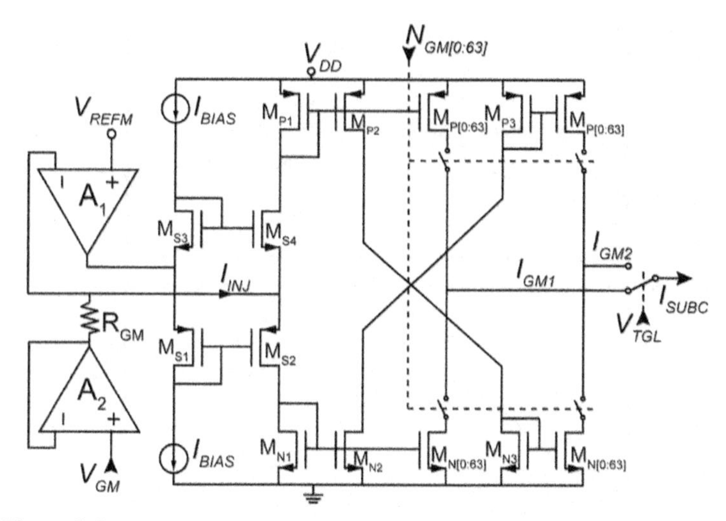

Figure 5.6 Push-pull transconductance amplifier (G_M) schematic diagram.

DAC (instead of varying R_{GM}) because it does not change the bias voltage in the GM circuit. Output current I_{SUBC} has a range of 4–15 μA in magnitude. It modulates the transmitter ICO, which translates into a range of 0.25–1 GHz for the FM-UWB signal bandwidth. The bandwidth is calibrated by setting V_{GM} equal to V_{REFH} (internal reference voltage of 0.5 V) and employing the SAR-FLL calibration scheme described in Section 3.2.5 to adjust N_{GM}.

There are two current DACs (one is a mirror of the other) generating anti-phase currents I_{GM1} and I_{GM2}. Selecting one of the DACs generates a triangle output current waveshape. Alternating both DACs synchronously using V_{TGL} generates a sawtooth output current, as illustrated in Figure 5.7. The sawtooth current waveform adds the possibility of modulating the sub-carrier wave using phase-shift keying (see Figure 2.9 for PSK modulation scheme).

Amplifiers A_1 and A_2 use a 2-stage folded-cascode topology and have 70-dB DC gain and 20-MHz unity-gain bandwidth. The wide bandwidth is required to reproduce a triangle wave with the desired fidelity (i.e., preserving up to the 21st harmonic for a 1-MHz sub-carrier). All current mirrors and sources shown in Figure 5.6 are cascoded and gain boosted to minimize current variation due to supply voltage changes. The G_M block has an effective transconductance of 8.4 μA/V and consumes 30 μA from a 1-V supply.

5.3.4 Regenerative Amplifier and Bandpass Filter

The noise figure of the first amplifier in the receive path must be less than 7 dB to obtain an overall sensitivity better than −80 dBm. However, decreasing the

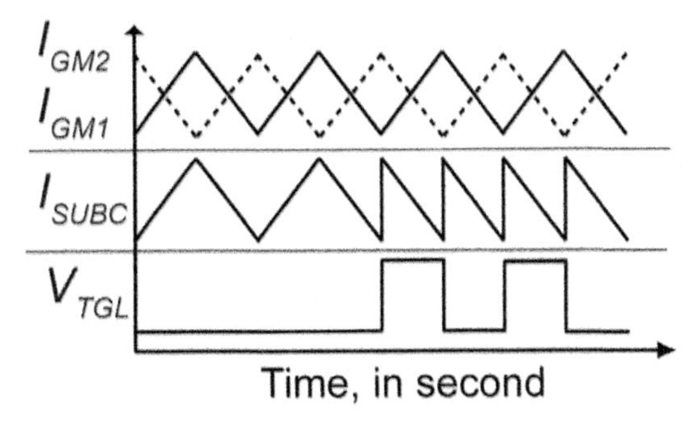

Figure 5.7 Timing diagram of triangle and sawtooth current modulations.

noise figure requires higher power consumption, in general [3]. Moreover, the amplifier gain should be greater than 30 dB so that the received signal amplitude is sufficient to drive the envelope detector. An LC bandpass filter at the amplifier output (i.e., tuned load) converts the received FM-UWB signal into an AM signal. Positive feedback enhances the Q of the LC filter thereby increasing its voltage gain and shrinking the bandwidth to the desired value of 50 MHz, which increases the selectivity and sensitivity of the receiver [4, 5]. However, the feedback needs to be controlled to avoid oscillation. A calibration circuit that addresses this challenge will be described in the Section 5.3.7. Regenerative amplifiers have non-linear voltage gain; the highest gain is achieved when the input signal is the weakest [6].

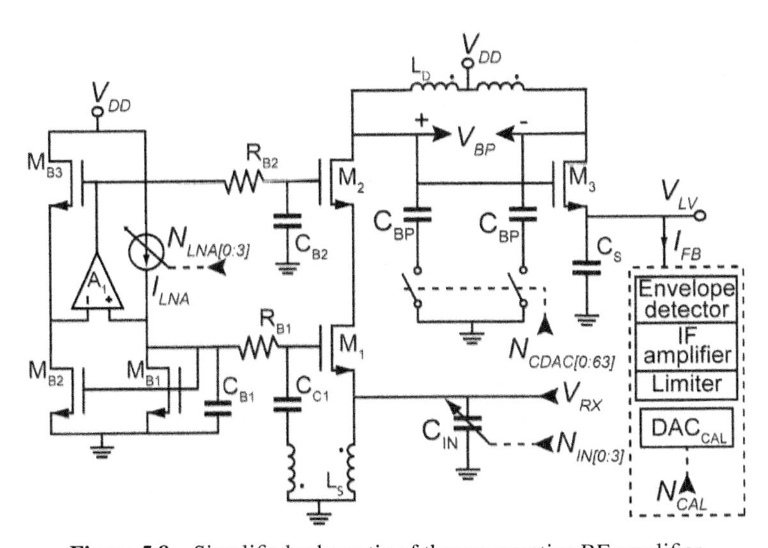

Figure 5.8 Simplified schematic of the regenerative RF amplifier.

The regenerative amplifier schematic is shown in Figure 5.8, where M_1 and M_2 form a cascode transconductor that drives an LC tank formed by L_D, C_{BP}, the gate capacitance of M_3, and parasitic capacitances at the drain of M_2. Cascode transistor M_2 increases the isolation between the RF input (V_{RX}) and output (V_{BP}) of the amplifier, while transistor M_3 forms a positive feedback loop at the preamplifier output via coupling between the 2 halves of transformer L_D. The amplifier gain increases quickly as the loop gain approaches unity, but its bandwidth also decreases as the gain-bandwidth product is approximately constant. The loop gain is given by

$$A\beta = kQ\omega L_D gm_3 \leq 1, \tag{5.2}$$

where k and Q are the coupling coefficient and quality factor, respectively, of inductor L_D. The transconductance of M_3 is gm_3, and ω is radian frequency. The loop gain is frequency dependent and depends mainly on fixed parameters of the passive inductor (L_D), so the loop gain can only be varied electronically by changing gm_3. The voltage gain of the overall regenerative amplifier is given by

$$A_V = \frac{V_{BP}}{V_{RX}} = \frac{gm_1 Z_O}{1 - A\beta} = \frac{gm_1 Z_O}{(1 - kQ\omega L_D gm_3)}, \tag{5.3}$$

where gm_1 is transconductance of M_1, and Z_O is impedance seen at the drain of M_2. It is clear from Equation (5.3) that the voltage gain of the amplifier is enhanced by increasing loop gain $A\beta$ (i.e., $A\beta \rightarrow 1$). The 5-pF capacitor C_S forces the source of transistor M_3 to ground at RF. DC current at the source of M_3 (I_{FB}) controls transconductance gm_3, thereby varying the amount of positive feedback. The source voltage of M_3 (V_{LV}, at approximately 0.6 V) is used as the supply voltage for other blocks (i.e., the envelope detector, IF amplifier and limiter). Bias current I_{FB} shown in Figure 5.8 consists of the DC current drawn by those 3 blocks plus the current drawn or injected by DAC_{CAL}, which is controlled by control word N_{CAL}.

Inductor L_D is 3.9 nH, with a self-resonant frequency of 9.1 GHz and peak-Q of 19.3 at 4.1 GHz. The value of C_{BP} is determined by a 6-bit capacitor DAC, and can range from 80 fF to 400 fF. The 6-bit code word N_{CDAC} tunes the center frequency of the load resonator in a range from 3.2–4.9 GHz. The DAC is thermometer coded, where a row and a column decoder connect V_{BP} to a set of capacitors, from the first capacitor (number 1) to the selected one (e.g., number 28) as illustrated in Figure 5.9(a). The capacitor DAC is not scaled uniformly, instead it scales progressively larger so that there is a linear relationship between the center frequency and word code, and a consistent

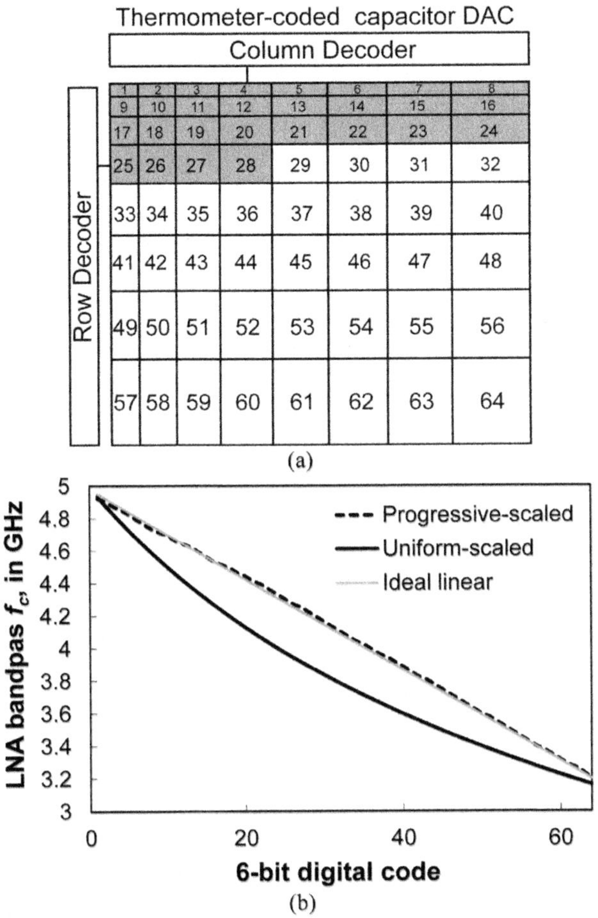

Figure 5.9 (a) Progressively-scaled capacitor DAC illustration and (b) simulation result compared to a uniformly-scaled capacitor DAC.

frequency step as the word is incremented. A comparison between progressive and uniform scaling from simulation is shown in Figure 5.9(b).

The preamplifier input is matched to 50 Ω across the 3–5 GHz band using matching network C_{IN} and L_S. The matched condition ($|S_{11}| < -10\,\text{dB}$) is achieved in several sub-bands (with overlap) by varying C_{IN} via 2-bit digital word N_{IN}, which controls the setting of a 2-bit capacitor DAC (see Figure 5.10). The sub-banded input matching automatically realizes minimum noise figure for the preamplifier in each sub-band frequency range, because the signal gain (passive gain due to the input tank) is highest at the frequency where impedance matching occurs. Inductor (L_S) is 3.6 nH, with a self-resonant

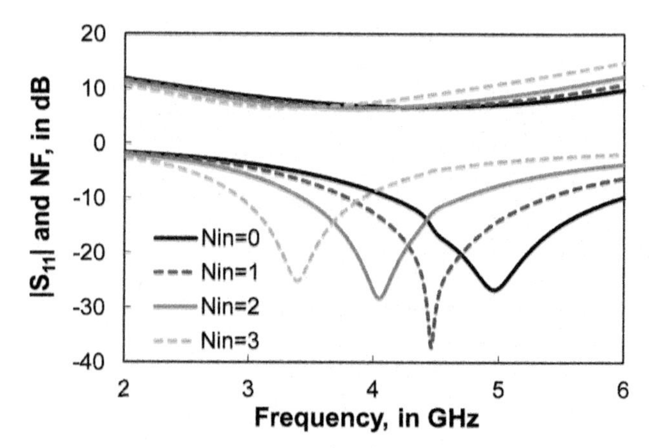

Figure 5.10 Simulation results of the $|S_{11}|$ and noise figure for various N_{IN}.

frequency of 9.6 GHz and peak-Q of 20.2 at 4.2 GHz. Both L_D and L_S are fully-symmetric on-chip inductors that provide the highest Q-factor in the smallest possible chip area [7]. Shunt-series negative feedback at the input of the main amplifier is also provided by L_S, which increases the input bandwidth from 0.45 GHz to 1 GHz. The opposite phase of the received signal appears at the gate of M_1 via 3-pF coupling capacitor C_{C1}, effectively doubling the input signal amplitude. The larger input signal lowers the 50-Ω noise figure from 9.1 dB ($C_{C1} = 0$) to 6.6 dB ($C_{C1} = 3$ pF). It should be noted that the main noise contribution comes from the (positive feedback) transistor M_3 (28%) and the main amplifier transistor M_1 (24%) when the RF input frequency is 4 GHz.

The amplifier is biased using a feedback-controlled replica circuit (A_1 and M_{B1}–M_{B3} shown in Figure 5.8) so that the bias is made insensitive to supply voltage variation without loading the drain of M_1. The measured bias current of the preamplifier varies by just ±1.5% for a supply range of 0.9–1.1 V. A PTAT current biases the amplifier such that the transconductance of the main amplifier is desensitized to temperature changes. Bias current I_{LNA} can be halved or doubled via a 2-bit current DAC controlled by N_{LNA} to raise the receiver sensitivity, or to work in a low-power mode. In the high-sensitivity mode, the amplifier noise figure improves, suitable when a higher SNR is required at the expense of higher power consumption. On another hand, the noise figure is poorer in the low-power mode, which may be suitable when there is plenty of link margin (e.g., communication across a shorter range).

The regenerative amplifier has voltage gain of 40 dB when it is tuned for optimal positive feedback at a loop gain ($A\beta$) of 0.9. The regenerative amplifier consumes 0.56 mA from the 1-V supply (0.45 mA for the 1st stage, M_1 and M_2) at 4-GHz operating frequency, which is 75% less power than the FM-UWB amplifier design described in Chapter 4 for the same voltage gain.

5.3.5 Envelope Detector, IF-Amplifier, and Limiter

The envelope detector (see Figure 5.11) removes the carrier from the filtered RF signal. Transistors M_{NZ1} and M_{NZ2} are a differential half-wave rectifier which discharges the 2.7-pF capacitor C_{ENV} when the RF voltage swing peaks at input V_{BP}. Transistor M_P and passives R_P and C_P form an active load that gives a high impedance at the IF ($Z_L = ro_p$) and small impedance ($Z_L = 1/gm_p$) at DC. The DC component in the AM input signal after detection comes mainly from thermal noise and any narrowband interference. The envelope detector bias current (I_{BIAS}) is derived from a PVT-compensated current reference. Transistors M_{NZ1}, M_{NZ2}, and M_{NZ3} are matched, native threshold devices (typical V_{TH} is 40 mV), which enhance the sensitivity of the envelope detector (simulated input sensitivity is 1 mV_{p-p}). The envelope detector reuses bias current flowing through the regenerative amplifier, thereby saving power. It is biased at 40 μA from the DC voltage at the regenerative amplifier output (V_{LV} in Figure 5.8 and Figure 5.11), which ranges from 0.5–0.6 V.

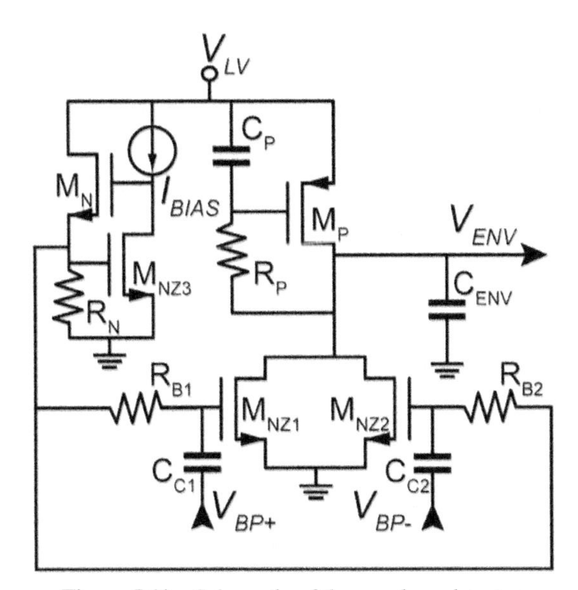

Figure 5.11 Schematic of the envelope detector.

The envelope detector is followed by an intermediate frequency (IF) amplifier and limiter, as shown in Figure 5.12. The 3-stage CMOS bandpass IF amplifier formed by transistors M_{N1}–M_{N3}, amplifies the detected envelope by 50 dB. It has a 100-kHz to 10-MHz bandwidth. Resistors R_{B1} and R_{B2} provide a DC feedback path for biasing the amplifier, while Miller capacitor C_M sharpens the upper roll-off of the bandpass filter [8]. Coupling capacitor C_{IF} provides bias isolation between the envelope detector and IF amplifier in the AC signal path. The output voltage of the IF amplifier (V_{IF}) is compared with the reference voltage generated by I_{REF} flowing through resistor R_{REF}, which is set by a 6-bit resistor DAC. Calibration via this DAC compensates for changes in the DC level of the V_{IF} signal and also for the offset voltage of the limiter with hysteresis.

Separate bias voltages for cascode devices $(V_{CSN1}–V_{CSN3}$ and $V_{CSP1}–V_{CSP3})$ are implemented to avoid creating a parasitic feedback path from the output to the input of the amplifier, where it may cause instability. A source follower buffer (not shown in Figure 5.12) is included so that V_{IF} may be measured for characterization and debugging purposes. A 2-bit DAC sets the bias current of the amplifier. For example, a higher bias current increases the upper cutoff frequency of the bandpass filter and raises the IF. The IF amplifier consumes 10–20 μA from source voltage V_{LV}, while the limiter consumes 5 μA from a 1-V supply at a 1-MHz IF.

5.3.6 FSK Demodulator

An FSK demodulator that detects the received data by demodulating the recovered sub-carrier IF signal is also implemented on-chip. As shown in Figure 5.13(a), the FSK modulated IF signal (V_{FSK}) is sampled by a clock

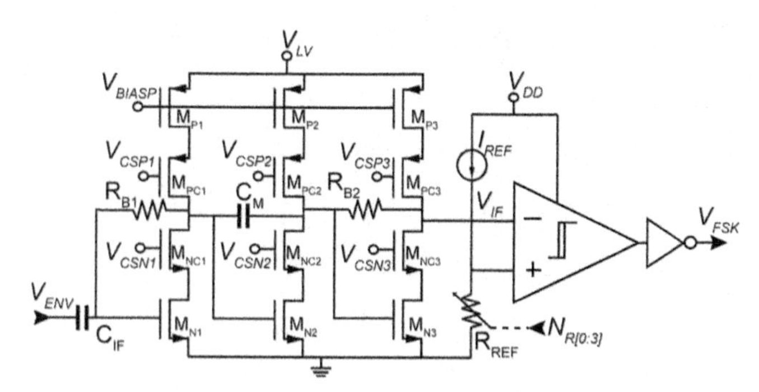

Figure 5.12 Schematic of the IF bandpass amplifier.

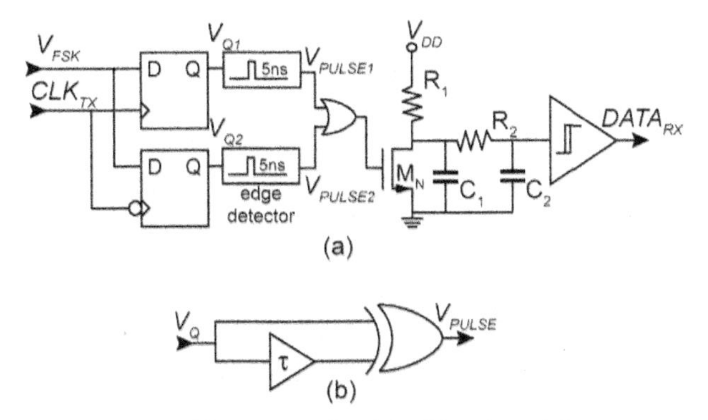

Figure 5.13 Schematic of the FSK demodulator.

(CLK_{TX}) sourced from the transmitter's sub-carrier oscillator using two edge-triggered D flip-flops. The sampled FSK signal (V_{BR}) produces a DC signal when V_{FSK} has the same frequency as CLK_{TX}, indicating a bit '0'. Otherwise, it produces a clock at a rate of 2Δf (typically 200 kHz) when the received bit is a '1'. Edge detector circuits are shown in Figure 5.13(b) and create a short duty cycle pulse at each of the flip-flop output edges. The pulse duration is 5 ns typically, but it is 3-bit programmable within a range of 2–20 ns by varying the delay time (τ) of the edge detector. Transistor M_N, which is driven by the pulses, drives low-pass filter R_{1-2} and C_{1-2} to average the pulse burst (in the case of a received '1'). The signal is then quantized by an inverter with hysteresis to recover the received data stream. The threshold of the inverter is adjustable digitally. The demodulator consumes an average of 4.6 μW from a 1-V supply at 1-MHz IF.

5.3.7 Positive Feedback Calibration

Automatic digital tuning is required to control the positive feedback loop of the regenerative amplifier. Simultaneous frequency and Q-tuning for the RF preamplifier can be supplied by a replica circuit [11]. However, this method suffers from mismatch, requires additional chip area, and consumes more power. The calibration method implemented here uses direct tuning [12], where the feedback is increased until oscillation occurs. In this way, the center frequency is detected and can be tuned using a digital FLL. After calibration of the center frequency, the loop gain is reduced (until it is less than unity) to suppress the oscillation. One drawback of this method is that data cannot be received during the calibration. However, the transceiver could be calibrated only when required (this requires supervision at the system level) to minimize this disadvantage of the direct tuning method.

A block diagram of the feedback calibration loop is shown in Figure 5.14(a). The total current of the regenerative amplifier (i.e., I_{FB}) is adjusted by a 10-bit current DAC, a 5-bit coarse control word ($N_{CAL-COARSE}$) with 2.4 µA per LSB and a 5-bit fine control word ($N_{CAL-FINE}$) with 0.08 µA per LSB. The $N_{CAL-COARSE}$ is set by an up/down counter. The $N_{CAL-FINE}$ with a range $+/-$ 1LSB of the coarse level is added to fine tune the positive feedback during data reception. Further optimization at system level is therefore possible using a preamble in the received data.

The up/down counter is driven by a clock signal from the sub-carrier oscillator through a frequency divider circuit that produces 1 ms/cycle, thus allowing enough time for oscillation to build up in each counter state. The timing diagram of the calibration cycle is shown in Figure 5.14(b). The DAC is zeroed by resetting the counter, and calibration is initiated by signal CAL_{PFB} at time t_1. The coarse current DAC is increased incrementally by the up/down counter until the envelope detector circuit (shown previously in Figure 5.11) and a comparator detect oscillation (i.e., at t_2 in Figure 5.14). Once oscillation has built up, the SAR-FLL circuitry performs a center frequency calibration

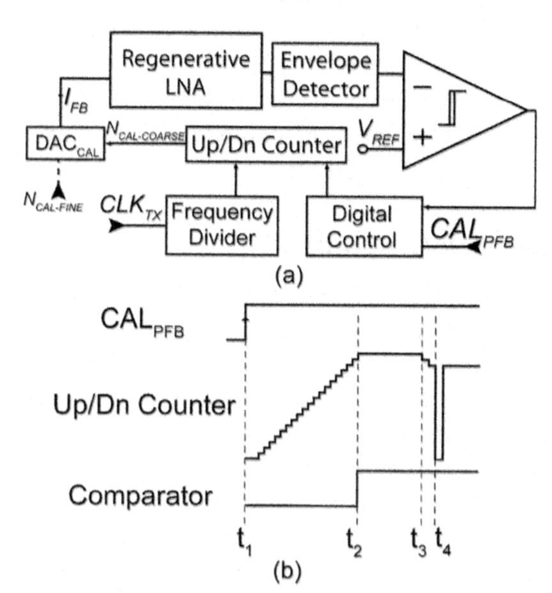

Figure 5.14 Block (a) and timing (b) diagrams for the positive feedback calibration loop.

(described in the following Section 5.3.8). After the frequency calibration is finished (at t_3), the counter is decremented for a few cycles to quench the oscillation. Simulations predict that two cycles (i.e., reducing I_{FB} by 4.8 μA and waiting for 2 ms) would set the amplifier at the optimal value where the RF gain-to-noise trade-off is optimized, and closed-loop gain $A\beta$ is approximately 0.9. A DAC reset is performed at t_4 for one clock cycle to quench any oscillation that might still occur due to hysteresis in the circuit.

5.3.8 Frequency Calibration

As the passive elements (R, L, and C) that define time constants on-chip vary in manufacture by (an estimated) ±10%, a frequency calibration scheme is needed to control the operating frequency or frequency range of various blocks. The successive approximation frequency-locked-loop (SAR-FLL) is employed to calibrate or directly control the operating frequency. A detailed schematic and description of the operating principle of the SAR-FLL were presented in Section 3.2.5. Figure 5.15 shows that the RF-ICO, sub-carrier oscillator, and the LC tank in the regenerative amplifier share the same

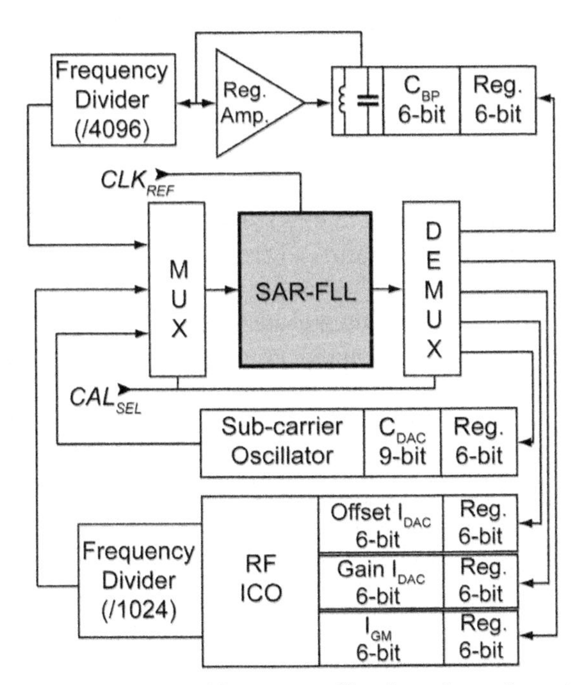

Figure 5.15 Block diagram of frequency calibration scheme for various blocks.

SAR-FLL calibration block through multiplexing (MUX) and de-multiplexing (DEMUX) selection blocks. The CAL_{SEL} control word selects a block that will be calibrated (i.e., selecting one path in the MUX and DEMUX), and the calibration result is stored in a 6-bit register. The RF-ICO and RF regenerative amplifier outputs are frequency divided to the IF range of 0.5–4 MHz for the calibration loop. A programmable reference clock (CLK_{REF}) with the same frequency range as the IF is assumed to be available externally (e.g., a crystal oscillator with a programmable synthesizer).

The regenerative amplifier (described in Section 5.3.4) center frequency is controlled by the 6-bit capacitive DAC (C_{BP}) through the SAR-FLL calibration loop when it is in oscillation mode. The oscillation frequency is divided by 4096 and compared to the CLK_{REF}. The calibration loop will set the C_{BP} so that the tank resonant frequency is equal to CLK_{REF} times 4096.

The sub-carrier oscillator frequency (described in Section 3.2.3) is controlled by a 9-bit C_{DAC}. The 3 MSBs are manually controlled to select various bands of sub-carrier frequency. The 6 LSB bits are set by the SAR-FLL such that the sub-carrier frequency (IF) is equal to CLK_{REF}.

The RF-ICO I_{DAC} (see Section 5.3.1) that controls the RF is calibrated such that its frequency range fits the desired operating range. The SAR-FLL calibrates the offset and the slope/gain of the transfer function between the center frequency and the DAC control word (N_{ICO}). The SAR-FLL calibrates 3-GHz and 5-GHz center frequencies at $N_{ICO} = 0$ and 63, respectively. The FM bandwidth is calibrated by controlling N_{GM} in the GM block (see Section 5.3.3) when input V_{GM} is set to V_{REFH} at 0.5 V. The carrier frequency is shifted from the center to a lower value, where the difference will be one-half of the desired FM bandwidth.

A tracking loop can be used to compensate for any change in the operating frequency due to temperature or supply changes without interrupting the transmission [13]. For example, the RF-ICO frequency can be monitored during modulation, as its average is approximately the same as its center frequency without modulation. All the calibration DAC control bits can be changed manually for characterization purposes and therefore it is possible to use other calibration schemes beside the SAR-FLL. In contrast with successive approximation, where the operating frequency hops around to find the best approximation, the frequency in the tracking loop is only changed step by step. The SAR scheme could be used in conjunction with the tracking loop running periodically in the background.

5.4 Measurement Results

A photomicrograph of the 0.9-mm^2 (including bondpads) transceiver prototype is shown in Figure 5.16. Each pad is ESD protected using a double diode cell from the standard-cell library that offers human body model (HBM) protection of at least 2 kV [1]. Receiver and transmitter are separated as far as possible to minimize coupling between them, while the bias and digital circuitry are located between them. All measured results presented in this section are from a test die wirebonded to a custom printed-circuit board (PCB) shown in Figure 5.17. A pair of 200-pF RF coupling capacitors are inserted

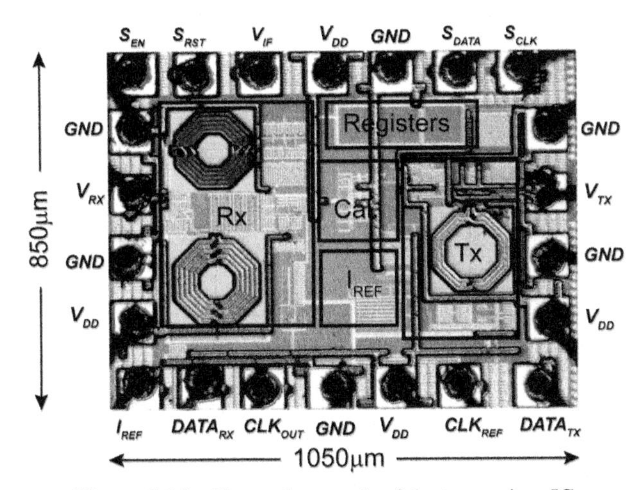

Figure 5.16 Photomicrograph of the transceiver IC.

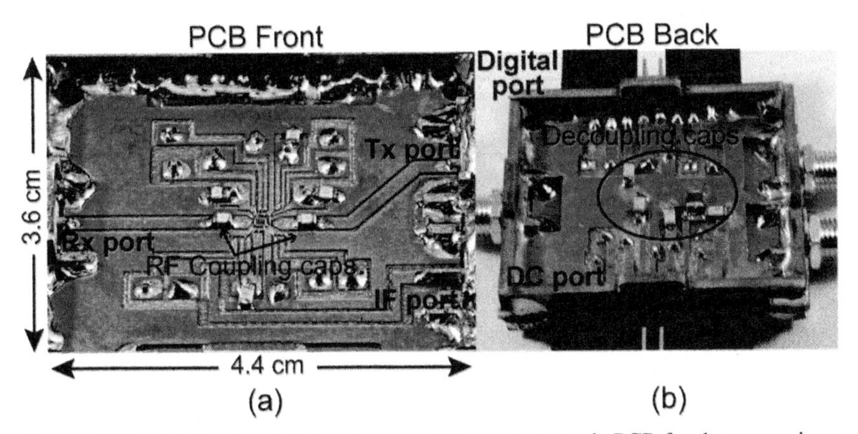

Figure 5.17 (a) Front and (b) back view of the custom made PCB for the transceiver.

in the V_{RX} and V_{TX} path to the connector. A set of capacitors (total value of 0.34 μF) are added to decouple the supply voltage (V_{DD}). The bondwire on the V_{TX} pad is placed manually so that it has a length of approximately 4 mm to realize a 4 nH inductor in the output matching network. Figure 5.18 shows the measurement setup, where the transceiver is controlled by a Matlab™ script through a field-programmable gate array (FPGA). The FPGA provides a serial digital interface to communicate with the test chip. The transceiver performance is characterized using a network analyzer, spectrum analyzer, oscilloscope, multimeter, and BER counter / pattern generator.

The transmitted center frequency and internal reference current are measured against supply voltage and temperature changes. There should be a correlation between the change in the reference current and the output

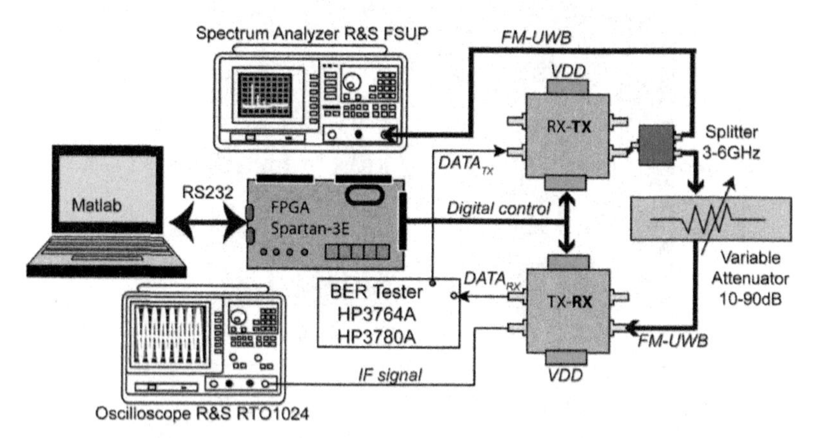

Figure 5.18 Measurement setup for the FM-UWB transceiver.

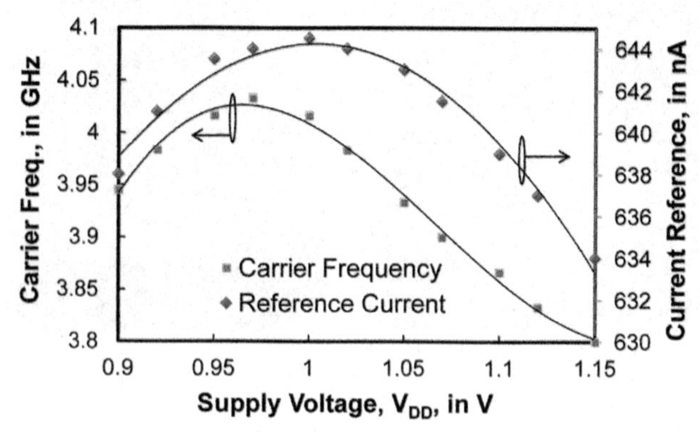

Figure 5.19 Measured reference current and carrier frequency versus supply voltage.

frequency generated by the RF-ICO. Figure 5.19 shows the variation of the reference current and transmitted carrier frequency versus supply voltage. The measured current has a parabolic shape with a peak located at a supply voltage of 1V. The dependence of the carrier frequency on supply voltage should track the reference current, however, it is shifted slightly. The carrier frequency varies by $\pm 2.5\%$ across supply voltage range of 0.9–1.1 V.

Figure 5.20 shows the variation of the reference current and carrier frequency versus temperature. The PCB was put on a hot plate and the temperature was measured using a thermocouple and multimeter. The carrier frequency falls as the temperature increases, with an average sensitivity of 0.75 MHz/°C. The carrier frequency is insensitive to temperatures between 80–100 °C, where the current is PTAT. The experiment proves that a PTAT current could be used to bias the RF-ICO to desensitize the RF-ICO to temperature.

The transmitter carrier frequency was also measured against the ICO current DAC control word as shown in Figure 5.21. The default setting for the offset and gain does not give the desired frequency range. However, once the offset and gain are calibrated (see Section 5.3.8), the range easily fits into the desired 3–5 GHz band. The DAC word and carrier frequency have a monotonic relationship, although there is a slight non-linearity. The non-linearity could be caused by mismatch in the DAC and a change in the oscillation amplitude. The integral non-linearity (INL) of the RF-ICO is ± 1.5 LSB (see Figure 5.22). The linearity is sufficient to give a precise frequency control and a flat FM-UWB spectral density.

The measured FM-UWB modulated signal bandwidth is plotted as a function of the GM DAC control word (N_{GM}) in Figure 5.23. The bandwidth can be controlled in the range of 265–960 MHz, and it is linearly related to the GM DAC current.

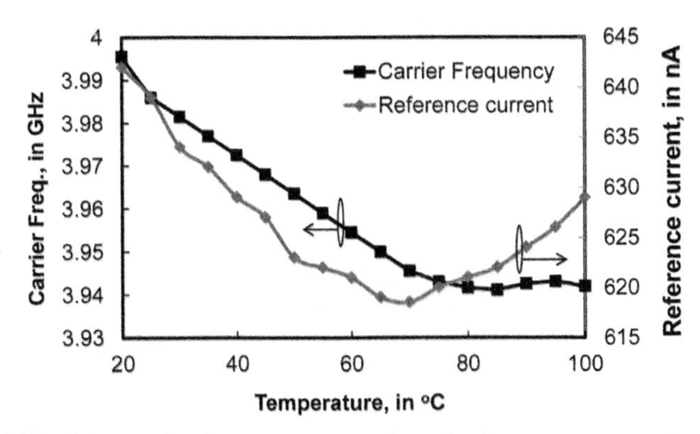

Figure 5.20 Measured reference current and carrier frequency versus temperature.

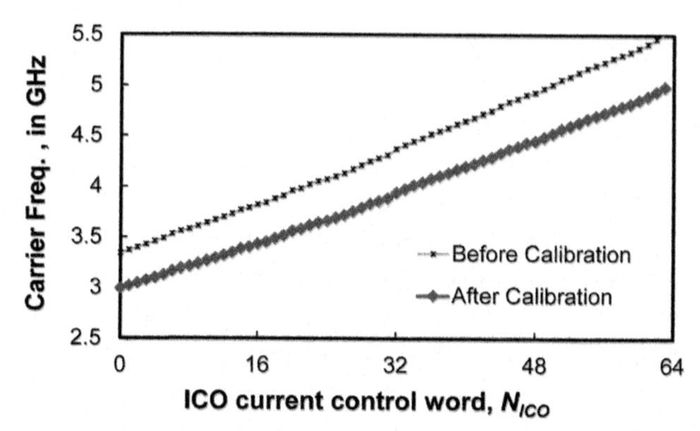

Figure 5.21 Measured carrier frequency versus DAC current controlled by N_{ICO}.

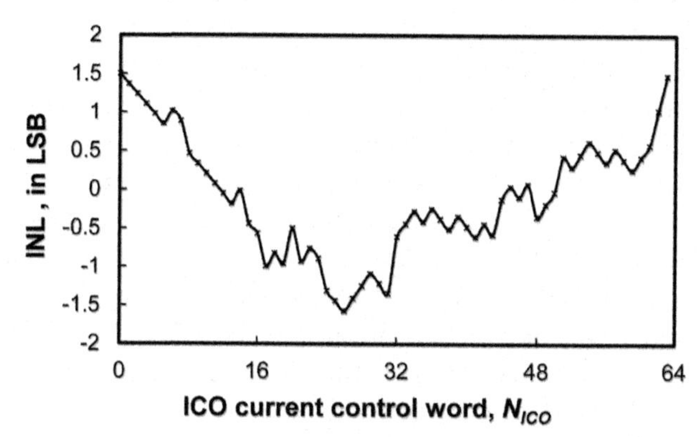

Figure 5.22 Integral non-linearity (INL) of the RF-ICO.

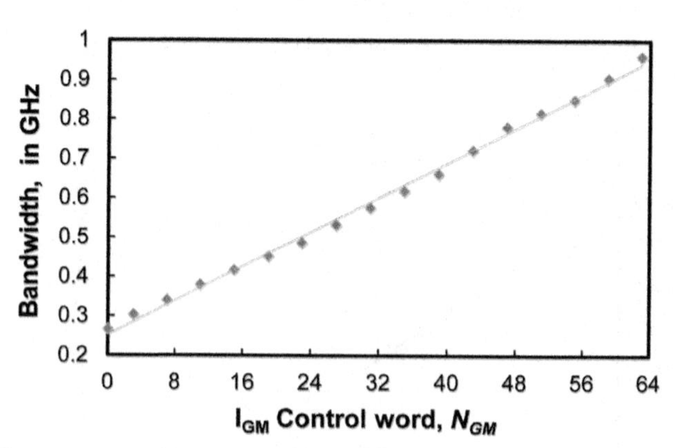

Figure 5.23 Measured FM bandwidth versus DAC current that is controlled by N_{GM}.

The spectral density of the modulated FM-UWB transmitter output for a center frequency of 4 GHz and bandwidth of 500 MHz is shown in Figure 5.24. The output power is adjusted so that it is just below the FCC indoor UWB mask, and it is measured in a resolution bandwidth of 1 MHz. The unmodulated carrier spectrum is shown in Figure 5.25, where the sub-harmonic component of the carrier can be seen. All harmonic components exist due to random mismatch in the 3-phase signals that are combined by the frequency tripler. The harmonic components are at least 30 dB lower than the main carrier, and they can be further suppressed by the bandpass response of the antenna.

The transmitter output power versus frequency measured in the 3–5 GHz range is plotted in Figure 5.26 for various output matching conditions (by changing N_{OUT}). Each N_{OUT} setting maximizes the transmitted output power

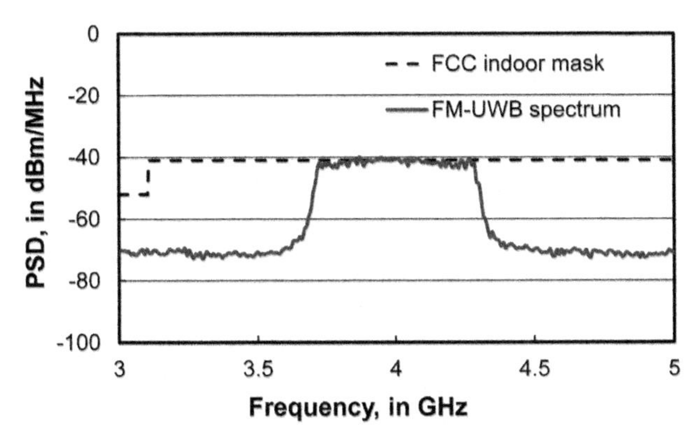

Figure 5.24 Measured FM-UWB modulated signal spectral density.

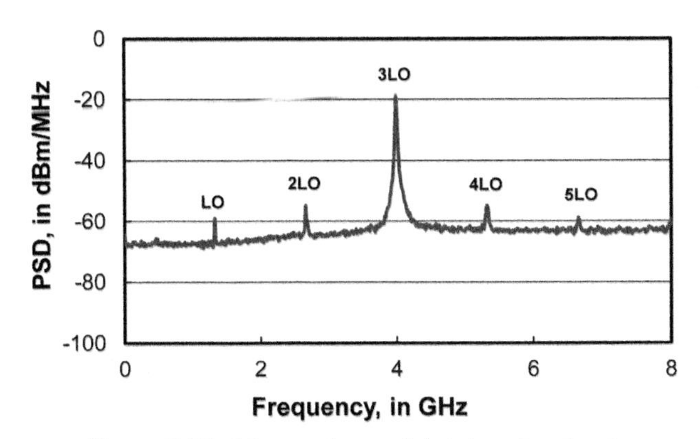

Figure 5.25 Measured unmodulated carrier signal.

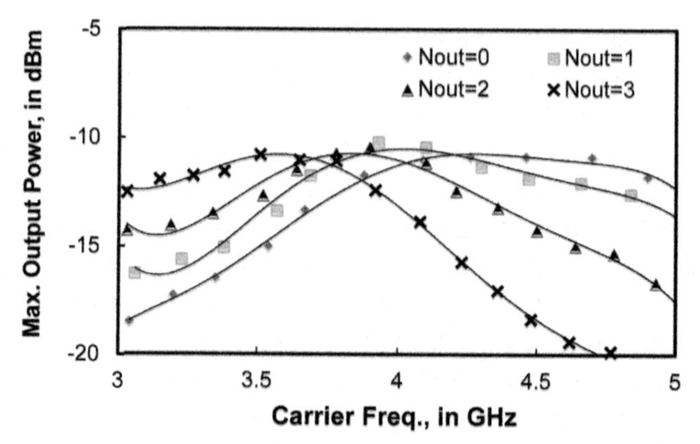

Figure 5.26 Measured maximum output power of the transmitter versus frequency at various N_{OUT}.

in a sub-band. The maximum output power at 4 GHz is −10.1 dBm, which is 1.5 dB lower than the expected value from simulation. The discrepancy could be caused by extra losses in the bondwire, PCB trace, and connector. The N_{PA} control word varies the output power in steps from −10.1, −11.2, −12.8, to −14.8 dBm, and the transmitter consumes 0.72, 0.67, 0.63, and 0.57 mW of DC power, respectively, at each output power level. An output power of −14 dBm is sufficient to transmit a 500-MHz bandwidth FM-UWB signal at the maximum allowable spectral density. Figure 5.27 shows the measured phase noise of the 4-GHz output carrier signal for an output power of −10.1 dBm. The phase noise is −70.7 dBc/Hz at 1-MHz offset. The phase

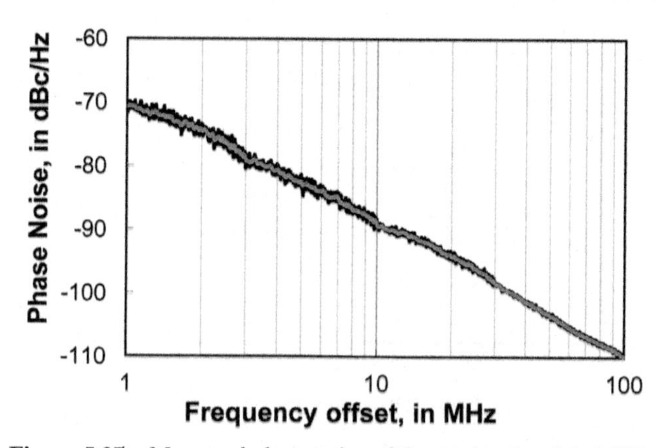

Figure 5.27 Measured phase noise of the carrier signal at 4 GHz.

noise result is comparable to the transmitter design described in Chapter 3. It meets the specification with 8.7-dB margin.

The receiver input impedance is reflected in the S_{11} measurement result shown in Figure 5.28. The result is close to the simulation result shown in Figure 5.10, where the receiver can be matched ($|S_{11}| < -10\,\text{dB}$) within a 1-GHz range. However, to cover the 3–5 GHz range, sub-band matching with overlap is desired. The results show that a tunable matching network is realized in CMOS technology.

The measured resonant frequency of the regenerative amplifier is shown in Figure 5.29. It is measured when the regenerative amplifier operates in

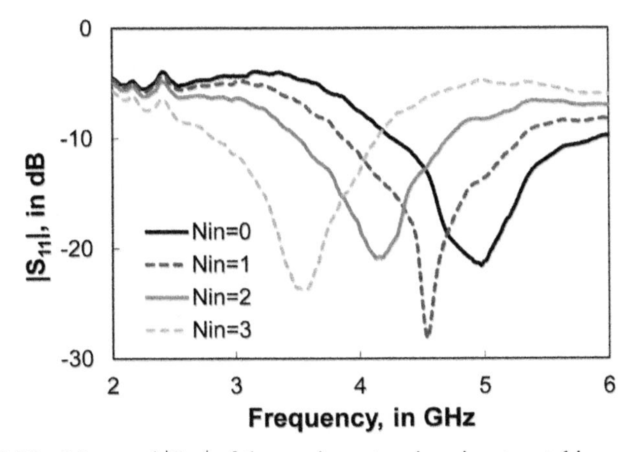

Figure 5.28 Measured $|S_{11}|$ of the receiver at various input matching conditions.

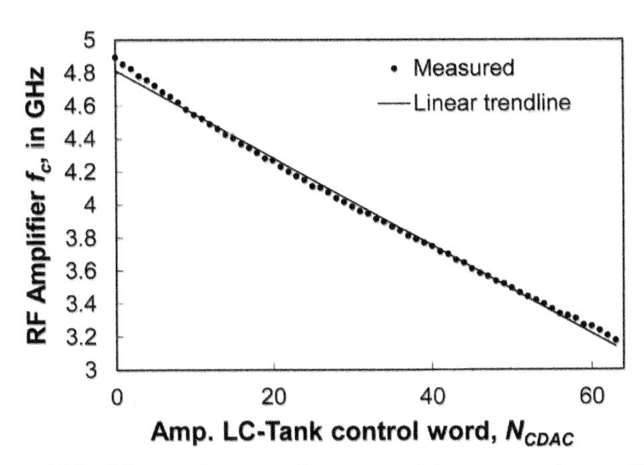

Figure 5.29 Measured resonant frequency of the regenerative amplifier.

oscillation mode and its output frequency is divided by 4096, and measured through the output clock buffer at CLK_{OUT}. Although progressive scaling is used, there is still non-linearity due to parasitics and processing variation in the capacitance values of the DAC.

The IF pulse generated by the envelope detector is measured via an on-chip buffer. The measured pulses at an IF of 3.2 MHz are shown in Figure 5.30 for an RF carrier of 4 GHz at a power of −70 dBm. It can be observed from the pulses that there is AM noise, which is removed by a limiter. A Fourier transform of the 1-MHz IF pulses is also shown in Figure 5.31 (2-FSK modulated at 100-kbit/s data rate). The frequency spacing between the tones is around 500 kHz in this case. It can be controlled by the relaxation

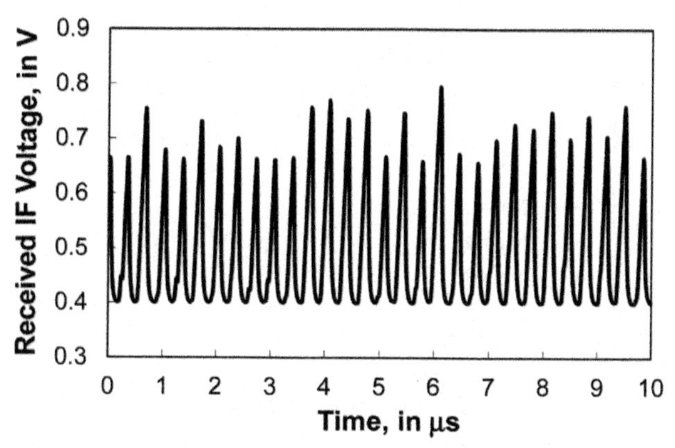

Figure 5.30 Measured IF pulses when the transceiver received −70 dBm FM-UWB signal.

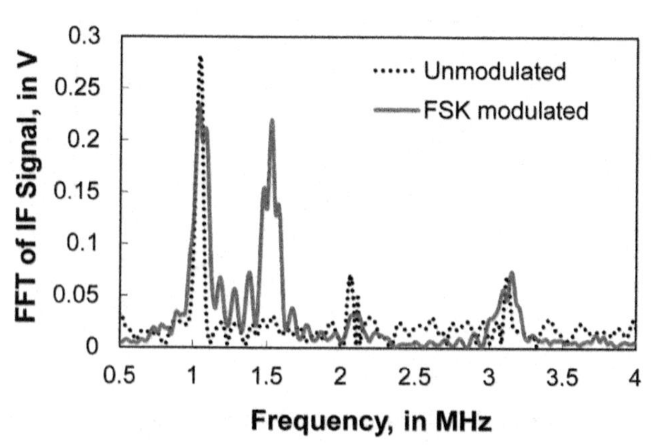

Figure 5.31 An FFT of the received IF signal with and without FSK modulation.

oscillator in the transmitter. The frequency spectrum of an unmodulated signal is also shown for comparison.

The FSK demodulator will detect the quantized IF pulse by sampling it using the same clock frequency (e.g., 1-MHz or 1.5-MHz clock for the IF in Figure 5.31). If the sampling rate and the IF are the same, then no pulse is generated, and vice versa. The measured pulse burst shown in Figure 5.32 is correlated (i.e., burst of pulses when data is '1', no pulses when data is '0') to the transmitted pseudo-random binary sequence (PRBS) data. The received bits for a data rate of 100 kbit/s and input power of −70 dBm are shown Figure 5.33. The received bits are identical to the transmitted bits,

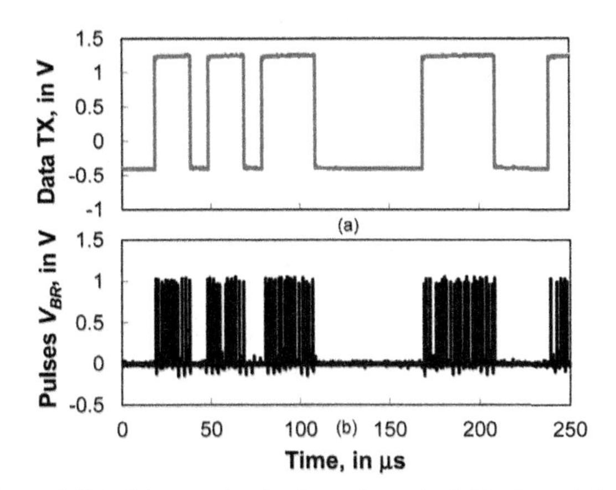

Figure 5.32 Measured pulse burst from the FSK demodulator.

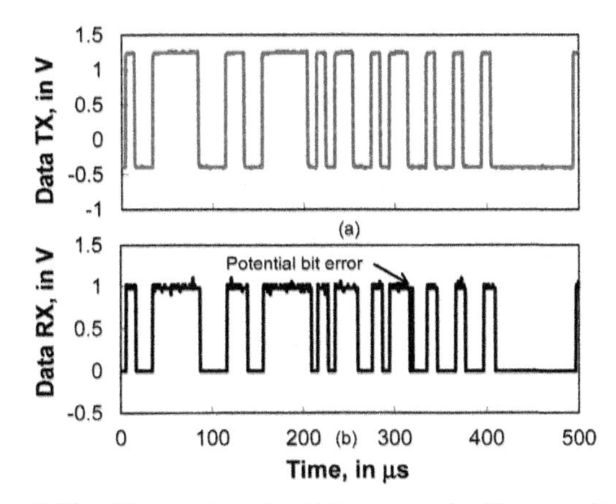

Figure 5.33 Measured received bit compared with transmitted bit.

but occasionally there is a glitch that could be interpreted as an error by the receiver. If the RF input signal power is reduced, the occurrence of the glitches increases.

The bit error rate for received data is also plotted against the received power in Figure 5.34 for a 4-GHz RF carrier under various conditions (i.e., different receiver DC powers and a higher data rate of 200 kbit/s). The regenerative amplifier bias can be controlled by the control word N_{LNA} to improve its noise figure at the expense of greater power consumption (see Figure 5.8). The noise figure directly affects the bit error rate (BER) and sensitivity (sensitivity is defined by input power when BER is 10^{-3}). The default setting (i.e., when the receiver consumes 0.58 mW) gives a sensitivity of −80.5 dBm, while the lower (0.43 mW) and higher (0.88 mW) power setting give sensitivities of −79 dBm and −82.5 dBm, respectively. The data rate in this receiver does not change the BER curve, however, at a data rate of 200 kbit/s the BER is limited to 5×10^{-2}. This is because of the limited bandwidth of the FSK demodulator. When an external demodulator is used, this constraint on the data rate is relaxed.

The sensitivity to single-tone interference is measured by detecting the degradation in BER as shown in Figure 5.35. The signal-to-interference ratio (SIR) measures the strength of the input signal compared to interference, and is negative when the interference power is stronger than input signal. The interference at frequencies of 2.4 and 5.7 GHz (ISM bands) gives a SIR limit (for BER of 10^{-3}) of −32 and −28 dB, respectively. In-band interference at 4 GHz can only be tolerated at a SIR level of −18 dB.

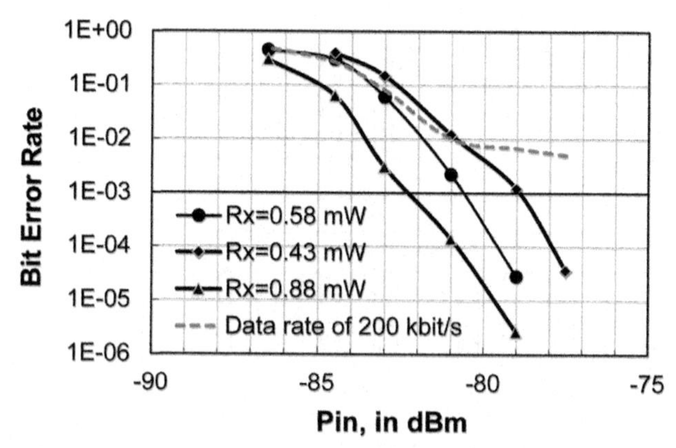

Figure 5.34 Measured BER versus received power for various Rx DC power consumption settings and data rates.

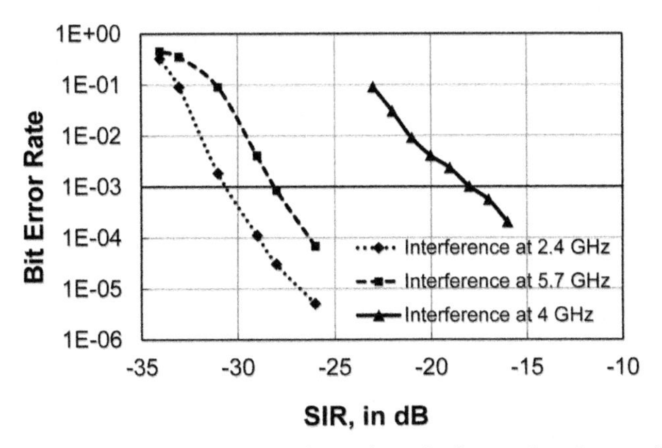

Figure 5.35 Measured BER versus SIR for various single-tone interference frequencies.

Figure 5.36 shows the power consumed by various building blocks for the FM-UWB transceiver in transmit and receive modes, respectively. The PA consumes the most DC power (78%) in the transmitter. Improving the PA can therefore directly benefit the overall power efficiency. On the receive side, the regenerative amplifier consumes the biggest portion of DC power (78%).

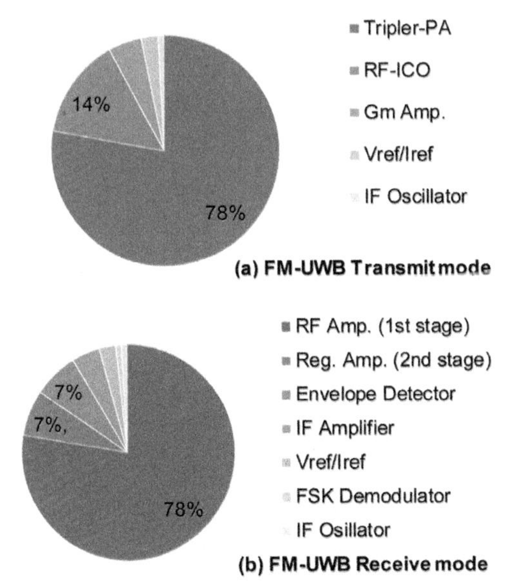

Figure 5.36 Power consumption distribution of the FM-UW transceiver for (a) transmit and (b) receive modes.

Table 5.1 shows a summary of the transceiver performance compared to recent FM-UWB transceivers published in the literature. This work is the first full transceiver that includes back-end processing in CMOS technology. The transceiver is compact and consumes only 0.55 mm^2 of active area, and achieves an average (Tx and Rx) energy efficiency of 6 nJ/bit at a data rate of 100 kbit/s. The transceiver is digitally controllable and consumes only 7 μW when put into standby mode (i.e., part of the bias circuit is turned on to get less than 1 μs wake-up time). The transmitter efficiency is 13.3%, which is an improvement from previous FM-UWB transmitter in Chapter 3. The measured receiver sensitivity and in-band SIR are −80.5 dBm and −18 dB, respectively. The performance meets the desired specification but is poorer compared to the previous receiver in Chapter 4. The degradation in sensitivity is caused by the non-optimal on-chip FSK demodulator. Sensitivity is improved by 3.4 dB if the IF signal is demodulated using an external demodulator (described in Section 4.4). Overall, the fully-integrated transceiver is suitable for low power, low data rate wireless applications that rely on harvested energy to power the device.

Table 5.1 FM-UWB transceiver performance summary

Parameters	This Work	[14]	[15]	[2, 8]
Technology	CMOS 90 nm	CMOS 180 nm	CMOS 130nm, SiGe BiCMOS	CMOS 90 & 65 nm
RF range, in GHz	3–5	3.43–4.27	7.2–7.7	4–5
V_{DD}, in V	1	1.6	1.1, 2.5	1
Phase noise, in dBc/Hz at 1 MHz	−71	−90	−107	−75
Max. Output Power, in dBm	−10.1	−13.7	−10	−10.2
Bandwidth, in MHz	265–960	560	500	500
Sub-carrier Frequency, in MHz	0.5–4.3	13.2	1–2	0.8
Receiver sensitivity at BER = 10^{-3}, in dBm	−80.5	−70	−87	−84
Signal to interference ratio, in dB	−18	–	−20	−30
Data rate, in kbit/s	100	50	100	100
Tx power consumption, in μW	630	8700	5000	900
Tx max. power efficiency, in %	13.3	0.5	2	9.1
Rx power consumption, in μW	580	7200	14000	2200
Standby power consumption, in μW	7	–	–	–
Active area, in mm^2	0.56	2.2	0.94	0.4
Tx energy efficiency, nJ/bit	6.3	174	50	9
Rx energy efficiency, nJ/bit	5.8	144	140	22

5.5 Conclusions

A fully-integrated FM-UWB transceiver (bits-in Tx/bits-out Rx) for the 3–5 GHz band which benchmarks the scheme for potential low-power, short-range applications has been realized in a 90-nm bulk CMOS technology. The 0.9-mm^2 full transceiver IC demonstrator consumes just 630 μW in transmit mode and 580 μW in receive mode from a 1-V supply. The transmitter generates a maximum RF output power of −10.1 dBm and achieves a power efficiency (RF output power divided by DC power) of 13.3%. The receiver achieves a sensitivity of −80.5 dBm at a BER of 10^{-3}, and in-band SIR of −18 dB for an NBI. On-chip calibration was implemented to tune the center frequencies of various blocks in the FM-UWB transceiver. The average energy consumption of the transceiver is 6 nJ/bit at a data rate of 100 kbit/s in continuous operation, with a standby power consumption of 7 μW. The transceiver is therefore suitable for low data rate, portable wireless applications powered from a battery or an energy harvester.

Table 5.2 compares the performance of the FM-UWB transceiver and other low-power transceivers from the literature. In term of energy efficiency, this work is comparable to other narrowband transceivers. FM-UWB has an advantage compared to other narrowband transceivers in terms of in-band SIR. Additionally, the FM receiver is generally more robust to amplitude noise compared to AM/OOK receiver. Moreover, the transceiver is tunable

Table 5.2 Low-power transceiver performance comparison

Parameters	This Work	[16]	[17]	[18]	[19]
Standard	FM-UWB	900 MHz ISM	Super-Regen.	2.4 GHz ISM	Super-Regen.
Technology	CMOS 90 nm	CMOS 180 nm	CMOS 180 nm	CMOS 130 nm	CMOS 180 nm
Modulation	2-FSK	OOK	OOK	BFSK	BFSK
RF band, in GHz	3–5	0.9	1.9	2.4	2.4
Power Consumption, in mW (Tx/Rx)	0.63/0.58	3.1/2.4	1.6/0.4	1.12/0.75	1.15/0.215
Receiver sensitivity, in dBm	−80.5	−71	−100.5	–	−86
Max. Output Power, in dBm	−10.1	−10	−4.4	−5	−5.2
Data rate, in kbit/s	100	1000	5	300	125
Signal to interference ratio, in dB	−18	–	–	–	–
Energy efficiency, in nJ/bit (Tx/Rx)	6.3/5.8	3.1/2.4	320/80	2.3/ 2.5	9.2/1.7
Active area, in mm^2	0.56	0.9	0.7	0.8	0.55

across a wide frequency range and has the ability to tune across the 3–5 GHz range to obtain the highest SNR. This work also has the smallest chip area among the other transceiver implementations, which is desirable for low-cost applications.

Further improvements can be made, especially in the FSK demodulator block, where it is seen to limit the sensitivity and data rate of the receiver. By employing more digital circuits and migrating to a smaller feature size (e.g., 28-nm CMOS), the chip area and power consumption can be reduced further while maintaining performance.

References

[1] B. Jagannathan, R. Groves, D. Goren, et al., "RF CMOS for microwave and mm-wave applications", *Proc. of Silicon Monolithic Integrated Circuits in RF Systems*, Jan. 2006, pp. 259–264.

[2] N. Saputra, J. R. Long, "A fully-integrated, short-range, low data rate FM-UWB transmitter in 90 nm CMOS," *IEEE Journal of Solid State Circuits*, Vol. 46, No. 7, pp. 1627–1635, July 2011.

[3] B. W. Cook, A. Molnar, K. S. J. Pister, "Low power RF design for sensor networks," I *proc. IEEE RFIC Symposium*, June 2005, pp. 357–360.

[4] D. Li, and Y. Tsividis, "Active LC filters on silicon," *IEEE Proc. Circuits Devices Systems*, Vol. 147, No. 1, Feb. 2000, pp. 49–56.

[5] W. B. Kuhn, F.W. Stephenson, A. Elshabini-Riad, "A 200 MHz CMOS Q-enhanced LC bandpass filter," *IEEE Journal of Solid State Circuits*, Vol. 31, No. 8, pp. 1112–1122, Aug. 1996.

[6] J. Tapson, T.J. Hamilton, C. Jin, A. van Schaik, "Self-tuned regenerative amplification and the hopf bifurcation," *Proc. of ISCAS*, Vol. 1, June 2008, pp. 1768–1771.

[7] M. Danesh and J. R. Long, "A differentially-driven symmetric microstrip inductor," *IEEE Transactions in Microwaves Theory and Techniques*, Vol. 50, No. 1, pp. 332–341, Jan. 2002.

[8] N. Saputra, J. R. Long, J. J. Pekarik, "A 2.2 mW regenerative FM-UWB receiver in 65 nm CMOS," *proc. IEEE RFIC Symposium*, May 2010, pp. 193–196.

[9] Yichuang Sun, *Design of High Frequency Integrated Analogue Filters*, The Institution of Electrical Engineers, 2002.

[10] J. Pandey, J. Shi, B. Otis, "A 120μW MICS/ISM-band FSK receiver with a 44μW low-power mode based on injection-locking and 9x frequency

multiplication," *IEEE Int. Solid-State Circuits Conf. Dig. Tech. Papers*, Feb. 2011, pp. 460–461.

[11] D. Li, and Y. P. Tsividis, "A loss-control feedback loop for VCO indirect tuning of RF integrated filters," *IEEE Transaction on Circuits and Systems-II: Analog and Digital Signal Processing*, Vol. 47, No. 3, pp. 169–175, March 2000.

[12] H. Ahmed, C. DeVries, R. Mason, "RF, Q-enhanced bandpass filters in standard 0.18μm CMOS with direct digital tuning," *Proc. of ISCAS*, Vol. 1, June 2003, pp. 577–580.

[13] Z. Zhao, A. Ivanov, "Embedded servo loop for ADC linearity testing," *Microelectronics Journal*, Vol. 33, pp. 773–780, 2002.

[14] B. Zhou, J. Qiao, R. He, et al., "A gated FM-UWB system with data-driven front-end power control," *IEEE Transactions on Circuits and Systems I*, Vol. 58, No. 12, Dec. 2011.

[15] J. F. M. Gerrits, H. Bonakdar, M. Detratti, et. al., "A 7.2–7.7 GHz FM-UWB transceiver prototype," proc. IEEE International Conference on Ultra-Wideband, pp. 580–585, Sept. 2009.

[16] D. C. Daly, A. P. Chandrakasan, "An energy efficient OOK transceiver for wireless sensor networks," *IEEE Journal of Solid State Circuits*, Vol. 42, No. 5, pp. 1003–1011, May 2007.

[17] B. Otis, Y. Chee, and J. Rabaey, "A 400μW-RX 1.6mW-TX super-regenerative transceiver for wireless sensor networks," in *IEEE ISSCC Dig. Tech. Papers*, Feb. 2005, pp. 396–397, 606.

[18] B. W. Cook, A. Berny, A. Molnar, S. Lanzisera, K. S. J. Pister, "Low-power 2.4-GHz transceiver with passive RX front-End and 400-mV Supply," *IEEE journal of solid state circuits*, Vol. 41, No. 12, pp. 2757–2766, Dec. 2006.

[19] J. Ayers, N. Panitantum, K. Mayaram, T. S. Fiez, "A 2.4GHz wireless transceiver with 0.95nJ/b link energy for multi-hop battery free wireless sensor networks," *Symp. on VLSI circuits dig. tech. papers*, June 2010, pp. 29–30.

6

Power Management for FM-UWB

6.1 Introduction

Wireless devices in an autonomous wireless network require electrical energy to operate and communicate with each other. Conventional approaches using energy stored in a battery limit the time when wireless communication takes place and the operational lifetime of the device. For unlimited operational time in an autonomous system, it is essential to have energy sources that can be harvested regularly and indefinitely. Research into energy harvesting has increased rapidly in recent years as indicated by the number of publications that contain the keywords "energy harvesting" (1500 publications in year 2015), listed by the IEEE-Xplore^{TM} website (see Figure 6.1).

Energy source can be light, thermal, vibration, and ambient RF, as listed in Table 6.1. In this work, a photovoltaic or solar cell is chosen because it generates the most energy indoors for a surface area of just a few cm^2. Furthermore, recent research results have demonstrated that it is also feasible to use a solar cell as the substrate for an antenna in wireless applications, thereby realizing a compact system [1–4]. A silicon solar cell achieves a typical efficiency of 10% and power density of 10 mW/cm^2 on a bright sunny day (e.g., irradiated at 1000 Watt/m^2) [5]. However, less than 1% of its maximum output power can be realized using indoor illumination. A $2 \times 2\,\text{cm}^2$ solar antenna (solant) is employed as an energy source for this work [6].

Unfortunately, solar and other renewable energy sources are intermittent. Therefore, energy is not available from the solant at all times (e.g., in darkness) unless harvested energy is stored in quantities sufficient to meet the demands of a wireless sensor node at any time. Furthermore, if the system is not running continuously, the energy drawn by the sensor node from the power supply is intermittent as well. Electrical energy storage is therefore required to collect and store the harvested energy. Energy harvested from the solant can be stored in a supercapacitor or a rechargeable battery. A "normal" capacitor cannot store a sufficient amount of charge, and a battery cannot be charged/discharged fast

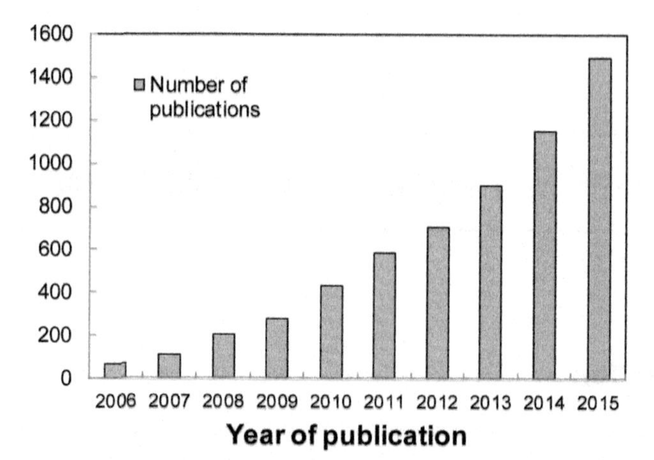

Figure 6.1 Numbers of publications that contain the keywords "energy harvesting" listed on the IEEEXplore™ website in recent years.

Table 6.1 Power estimates for microwatt energy harvesting [5]

Energy Source	Condition	Harvested Power Density
Vibration	Human	4 μW/cm^2
	Machine	100–800 μW/cm^2
Thermal (temperature difference)	Human	5–25 μW/cm^2
	Industrial	1–10 mW/cm^2
Light	Indoor	10–100 μW/cm^2
	Outdoor	10–25 mW/cm^2
Ambient RF	GSM	0.1 μW/cm^2
	WiFi	0.001 μW/cm^2

enough unless it occupies a volume much larger than a supercapacitor. A supercapacitor greater than 1 F can store enough energy from a solar antenna to level the energy demand from a sub-mW wireless sensor node in a wireless sensor network (WSN), a wireless medical body area network (WMBAN), for applications in the internet of things (IoT), etc., as outlined in Chapter 1.

The wireless front-end in an autonomous wireless system typically consumes most of the power overall [7]. A robust transceiver for autonomous applications utilizing FM-UWB modulation that consumes less than 1 mW has been described in Chapters 3–5 of this book [8, 9]. Other circuit functions, such as an on-chip sensing, memory, and a microcontroller may also be included, depending on the application. The supply voltage of a low-power CMOS transceiver is 1 V (typically), while circuits implemented in other technologies may have different supply voltage requirements.

Connecting the energy harvester to supply the wireless sensor system directly, (assuming that all circuits run from the same nominal supply voltage) yields poor efficiency, and is unreliable because the harvested supply voltage and current depend on the operating conditions of the solar cell (e.g., intensity of illumination). A power management sub-system is therefore required to regulate the DC supply to the autonomous wireless system, store harvested energy, and generate energy optimized interfaces for the transducers.

The objectives of this work are to design and demonstrate a power management sub-system that can make an autonomous wireless system implemented on a single chip using CMOS technology feasible. The power management scheme shown in Figure 6.2 consists mainly of switches and a DC-to-DC converter. Energy is drawn from the battery initially to bootstrap the system (i.e., when the supercapacitor is uncharged). Energy produced by the solar antenna can then be accumulated onto supercapacitor C_{SCAP} via the charge pump. This energy could be stored for a few days, but it will diminish slowly as there is a small leakage current in the supercapacitor itself. The DC-DC converter draws energy from the supercapacitor to operate the FM-UWB transceiver, sensor, and digital processing circuitry in normal operation (i.e., when enough energy has been stored). It converts a voltage ranging from 0.6–2.75 V (i.e., V_{SCAP}) to a regulated supply voltage of 1 V for use by the transceiver. Several DC-DC converters might be needed to supply other power consumers that require different supply voltages.

The technical background of building blocks in the system, namely, the solar cell, battery, supercapacitor, and DC-DC converter are described briefly in Section 6.2. The power management test chip and the design of each circuit

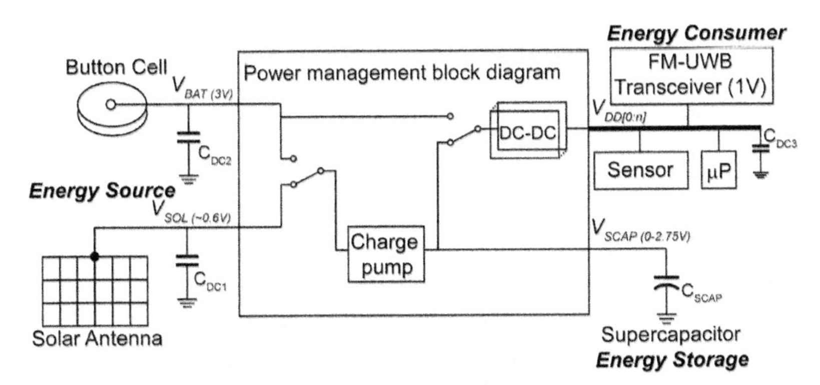

Figure 6.2 Block diagram of the proposed power management scheme for an autonomous wireless system.

block are detailed in Section 6.3. Experimental measurements of a prototype realized in 90-nm CMOS are then presented, followed by a brief comparison with other power management ICs reported from the literature in Section 6.4. Conclusions and suggestions for future work are summarized in Section 6.5.

6.2 Power Management Background

Batteries as an energy source as well as energy storage element are considered in this section. The solar cell as an energy source will also be described, along with its electrical model and characteristics. Supercapacitors used to store electrical charge are then described briefly. Finally, the advantages and disadvantages of various DC-DC converter circuits are discussed.

6.2.1 Solar Cell

The solar cell converts light energy from the sun to electricity. It is based on the photovoltaic effect, where a potential difference is generated at the junction of two different semiconductors in response to illumination by visible light (or other wavelengths). Free charge carriers are generated when photons are absorbed by materials forming a diode junction. Subsequently, the photo-generated charge carriers are separated and collected at the positive (anode) and negative (cathode) terminals of the cell. Photons within a specific range of frequencies, depending on the bandgap of the material, are absorbed and generate electron-hole pairs. Thus, different types of solar cell are sensitive to different wavelengths (λ) of light, for example, an amorphous-silicon (a-Si) solar cell is sensitive to wavelengths between 0.4–0.7 μm, while single-crystal silicon solar cells respond to 0.7–1.05 μm [10].

The solar cell is modeled as a diode and a current source, as shown in Figure 6.3(a). The electrical losses intrinsic to the solar cell are modeled by shunt and series resistors R_{SH} and R_{SE}, respectively. A typical I-V profile for a solar cell is illustrated in Figure 6.3(b). Maximum power output, P_{MAX}, from the solar cell is obtained at the knee of the curve, which corresponds to the optimum power voltage, V_{PMAX}. Short-circuit current (I_{SC}) flows though the external circuit when the electrodes of the solar cell are shorted. The open-circuit voltage (V_{OC}) is the terminal voltage when no current flows through the external circuit, which is the maximum voltage that a solar cell can deliver. V_{OC} corresponds to the voltage across a forward-biased p-n junction, and typically ranges from 400–900 mV depending upon the material used to fabricate the solar cell, and the light intensity [10]. As the energy of the arriving

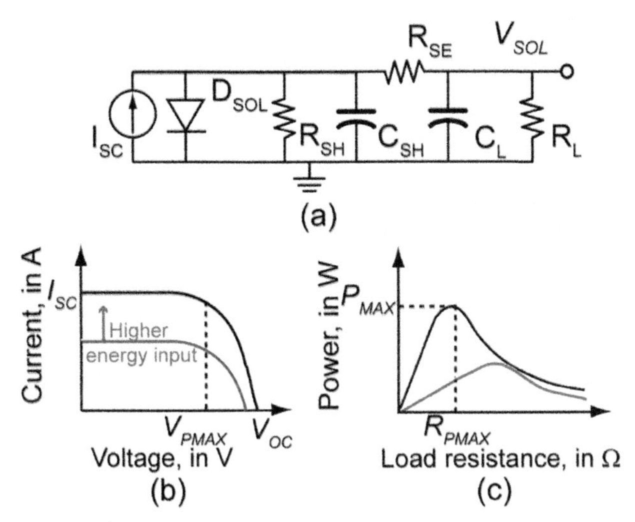

Figure 6.3 (a) Simplified electrical model of a solar cell, and (b) I-V relationship, and (c) power versus R_L for a solar cell.

photons increases, the I-V curve shifts upward, increasing I_{SC} dramatically while V_{OC} is only changed slightly. The fill factor is defined as the ratio of P_{MAX} to $I_{SC} \times V_{OC}$. It is a figure of merit indicating the quality of the solar cell. A typical fill factor value for amorphous silicon is 0.8 [10]. Figure 6.3(c) shows the output power versus load impedance, which indicates that there is an optimum load resistance which generates P_{MAX}. At a lower light input power, the optimum resistance shifts to a higher value. Therefore, an electronic interface that is attempting to collect energy from the solar cell as efficiently as possible under various lighting conditions should match the load and the cell impedances to obtain the maximum power transfer [11].

Conversion efficiency is defined as the ratio of P_{MAX} to the input power from a (standard) 1000 W/m^2 light source with air mass coefficient of 1.5 atmosphere thickness (AM1.5) [10]. Highest efficiency is obtained from GaAs solar cells, which offer efficiencies as high as 42% [12]. The amorphous silicon (a-Si) thin film solar cell used in the solant design for this work is 10% efficient (max.) [13]. Amorphous-silicon tends to have a lower efficiency compared to other solar cell materials, but it is cheaper to produce in large areas and quantities. However, due to the relatively high intrinsic source resistance R_{SH}, the efficiency of an a-Si solar cell degrades less in low light conditions (i.e., cell efficiency indoors relative to outdoors is around 90%) [14]. The efficiency-limiting factors for a solar cell are [15, 16]:

- Phonon generation (54%).
- Photons with energy not equal to the bandgap are not absorbed (18%).
- Usable energy lost by reflection (2%).
- Not all light is absorbed due to the limited thickness (2%).
- Shading due to the area used by electrodes (2%).
- Voltage drop due to series resistance R_{SE} of a solar cell or due to leakage current characterized by R_{SH} (6%).

6.2.2 Battery

The electrochemical battery is the conventional energy source for portable electronic devices. While IC technology has advanced by miniaturization and doubling of the transistor density every one and one-half years, battery technology does not benefit for such scaling, and energy density is advancing at a much slower rate, doubling every 10 years [5]. Batteries can be classified as non-rechargeable and rechargeable, and also by their electrochemical material composition. Table 6.2 is a short list of batteries available for portable applications at the time of writing. A lithium-ion (Li-ion) battery has an energy density of 200 Wh/kg. Assuming a 10-g Li-ion battery, a wireless transceiver that consumes 1 mW can operate for 3 months continuously. If

Table 6.2 Comparison of batteries suitable for portable devices [26]

Chemistry	Model	Voltage (V)	Capacity (mA-h)	Weight (g)	Volume (cm^3)	Energy Density (mW-h/cm^3)	Energy Density (mWh/g)
Alkaline	AAAA	1.5	625	20	1.20	781	47
NiMH	AAA	1.5	1000	13	3.85	389	115
	AA	1.5	2900	30	7.91	550	145
	C	1.5	6000	80	26.53	339	113
Li-ion	CR1225	3	50	2	0.28	531	75
	CR1632	3	120	5	0.64	560	72
	CR2032	3	225	2.9	1.00	672	233
	CR2450	3	610	6.9	2.08	880	265
	CR2477	3	1000	10	3.48	862	300
Silver Oxide	SR43	1.55	120	2	0.44	419	93
	SR44	1.55	175	2.5	0.57	476	109
	SR66	1.55	26	1	0.09	427	40
Li-ion	MS920S	3.1	11	0.47	0.15	229	73
Rechargeable	ML2032	3	65	3	1.00	194	65
NiMH	V600HR	1.2	600	14.5	4.29	168	50
Rechargeable	V20HR	1.2	20	1	0.21	116	24

10% duty cycling is used for the transceiver power supply, the operational lifetime is (ideally) increased to 2.5 years. However, the battery itself has a limited lifetime of around 5 to 10 years due to self-discharge caused by internal and external current leakages. Also, replacing a battery in the field may be impossible, or may increase the maintenance cost for an autonomous wireless system. Long-term, maintenance-free operation is preferred. Using harvested energy as an additional source to supplement or recharge the battery periodically can increase the lifetime and utility of an autonomous wireless system.

A fully-charged battery is modeled by a voltage source in series with a non-linear internal resistance [17]. A typical discharge curve of a battery for a constant load current is shown in Figure 6.4. The battery provides almost a constant DC output voltage in the range 0.9–3.2 V, depending on the type [18]. When the stored energy is depleted, the output voltage collapses suddenly. The charge capacity of the battery is sensitive to temperature and the magnitude of the load current. At loads above its rating, over-discharge occurs, and the battery is depleted faster. The internal resistance parasitic energy losses increase [19]. The limit on the current that can be pulled from a battery depends on the battery chemistry, volume, and construction.

The autonomous wireless transceiver is usually duty cycled, so the battery output current has a pulse like profile over time [20]. However, duty-cycling a radio can draw a transient over-discharge current from the battery, which has a detrimental effect on the battery lifetime. A capacitor connected in parallel with the battery supplies enough transient current to extend the battery lifetime by up to 1.7x under pulsed load conditions for load currents of just tens of mA [21]. It has also been shown that fast duty cycling (i.e., a few kHz) reduces

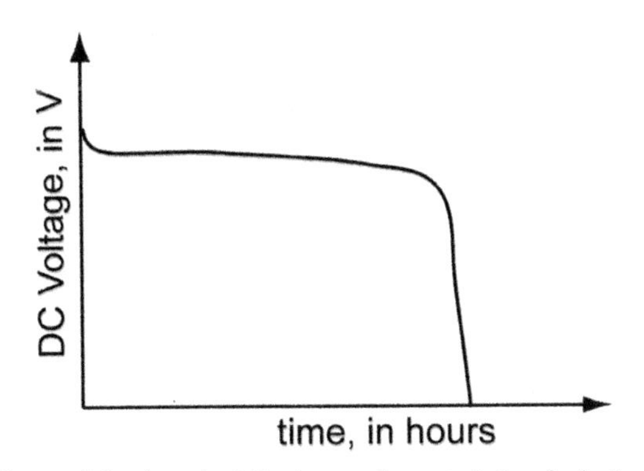

Figure 6.4 A typical discharge characteristic of a battery.

battery capacity by half compared to duty-cycling at a frequency ten times slower [22].

A rechargeable battery can also store harvested energy. Unfortunately, rechargeable batteries have a lower energy density and higher leakage current than non-rechargeable batteries [23]. The charging current must be controlled to avoid excessive heating or damage due to overcharging [24]. Limits on the number of recharge cycles (around few hundred times [25]) and the need for extra circuitry to manage the charging process add to the complexity of the power management system when rechargeable batteries are used. Compared to a capacitor, it is more difficult to monitor the amount of energy that is stored in a battery, as the terminal voltage is almost the same whether the battery is full or discharged (see Figure 6.4). The maximum voltage decreases as the number of charge cycles increases [25].

6.2.3 Supercapacitor

Unlike a battery, a capacitor is more tolerant to a wide range of energy transferred rates when charging or discharging (power density is in the range of 10^3–10^6 W/kg). Power is defined as the rate of energy transfer. High energy density elements can store more energy, while components with a high power density can be filled or emptied in a short time. Typically, high energy density storage cells have low power density and vice-versa (see Figure 6.5). The energy density of a ceramic or tantalum dielectric capacitor is much lower than a typical battery. Supercapacitors (a.k.a. ultracapacitors) have an energy density 100 times greater than a normal capacitor, as shown in Figure 6.5. The number of recharge cycles for a supercapacitor is approximately 1 million, which is very favorable compared to a battery, which may be recharged less than a thousand times.

Supercapacitors range from 0.01–10 F and have a packaging area of 1–4 cm^2. The leakage current of a supercapacitor ranges between 0.5–20 μA, and its effective series resistance (ESR) is typically less than 1 Ω [28]. It also has a maximum charge/discharge current rating on the order of few amperes [28]. Absolute maximum voltage ratings are approximately 2.5 V for a single supercapacitor, or 5 V for two capacitors connected in series [29].

The voltage across the supercapacitor, V_{SCAP}, depends on how much energy is stored, which can be expressed by

$$V_{CAP} = \sqrt{\frac{2E}{C_{SCAP}}}, \tag{6.1}$$

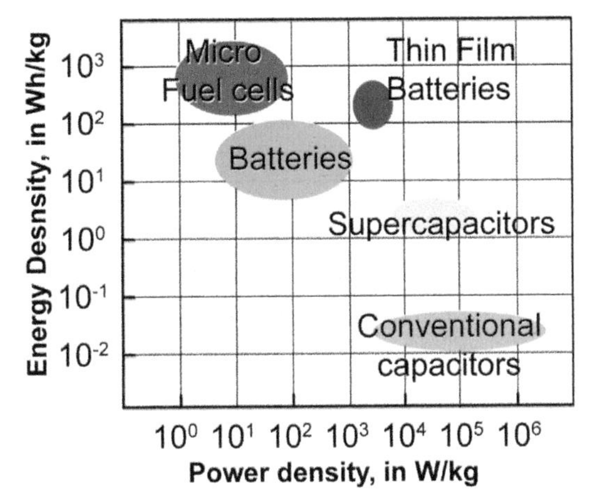

Figure 6.5 Energy versus power density for various energy storage methods [27].

where E is the energy stored, and C_{SCAP} is the supercapacitor size. Knowing the amount of energy that is stored by sensing the terminal voltage can be advantageous for power management purposes. However, a wide input range DC-DC converter is required to supply the load with because the terminal voltage of the capacitor will drop towards zero as it is discharged.

6.2.4 DC-DC Converter

The output voltage from an energy harvester must be regulated and voltage conversions are required between the source and load (i.e., the solar cell terminals and the wireless transceiver). There are two classes of DC-DC conversion techniques used commonly in power management sub-systems. The first is the linear voltage regulator, and the other is the switching regulator.

A linear regulator, as shown in Figure 6.6, provides regulation using a pass device (e.g., a transistor). Output voltage, V_{OUT}, is sensed by a resistive divider, and then fed back and compared with reference voltage, V_{REF}. The voltage difference is amplified by the error amplifier A_E that controls the pass device in a negative feedback loop. Feedback forces V_{OUT} to equal $(1 + R_{F1}/R_{F2})V_{REF}$ in the steady-state. The pass device controls the amount of current flowing from V_{IN} to V_{OUT}, and can be implemented using either NMOS or PMOS devices in a CMOS technology. An efficient regulator consumes little quiescent current, so most of the current feed the load through the pass device. The efficiency of the regulator can be estimated by the ratio of the output voltage divided by the input voltage, which implies that there

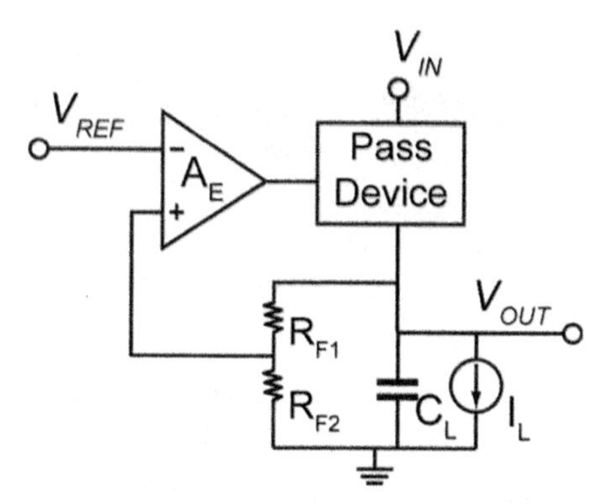

Figure 6.6 Schematic of a typical LDO.

is a power dissipated by the pass device. A low voltage drop across the pass device is desirable, and low dropout (LDO) voltage regulators are designed to minimize this voltage drop. The voltage drop cannot be zero, because a drain-source bias voltage is required to ensure that the CMOS pass device works in the saturation region. As V_{OUT} decreases, the voltage dropped across the pass device becomes a more significant part of the loss, and efficiency suffers.

The LDO has a low impedance at the regulated output at low frequency due to negative feedback, and at high-frequency due to shunt decoupling capacitor, C_L. Its simple implementation and compact size make the LDO a popular choice for a DC-DC converter. However, DC voltage step-up between input and output is not possible, because the output voltage must be always lower than the input voltage for proper operation.

Another type of a DC-DC converter is the switching regulator, which transfers a small amount of energy at a regular rate from the input source to the output. This is accomplished with the help of electrical switches and a controller to regulate the rate of the energy transfer. Switching regulators offer higher efficiency than linear regulators because energy is transferred to the load via either a capacitor or an inductor, which has low energy losses. Inefficiency results from parasitic losses in the switches and switch drivers. The switching regulator allows either a step-up (boost) or step-down (buck) of the DC voltage between its input and output. Unfortunately, it has the disadvantage of generating unwanted periodic voltage variations at the output called ripple. The ripple voltage (if large enough in amplitude) can generate in-band interference in a radio transceiver, or reduce the resolution of an ADC or DAC. The passive component values used in the switching regulator are inversely proportional to the clock frequency, and proportional to the load current required. This may result in a large chip area in contrast with a linear

LDO, which can be made more compactly. Fortunately, the 1 mA load current required from the on-chip DC-DC converter in this work can be implemented within a reasonable chip area ($< 1\,\text{mm}^2$).

Switching regulators can be divided into two types depending on the passive device that transfers energy. The buck and boost (step-down and step-up) switched-inductor topology shown in Figure 6.7 charges the output voltage by inductive current. The switch controls the current flow through the inductor by changing the duty cycle of the clock signal driving the switch. Diode D_1 ensures that current cannot flow back from the load towards source V_{IN}. In CMOS technology, the switches and diodes are usually implemented using MOSFETs [30]. A switched-inductor topology is able to achieve a peak efficiency higher than 90% [30]. Step-up or step-down DC conversion requires different L-C network configurations, therefore this type of switching regulator is not easily reconfigured from step-up to step-down on the fly (see Figure 6.7). A low ($< 0.5\Omega$) equivalent series resistance (ESR) inductor is required, which must be implemented off-chip [32]. Inductive (voltage) switching transients are a strong source of electromagnetic interference (EMI) that can disturb a wireless transceiver or a sensor.

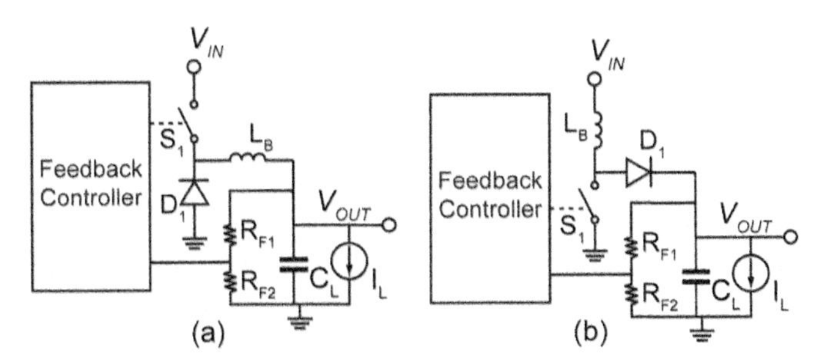

Figure 6.7 Schematic of (a) buck and (b) boost switched-inductor DC-DC converters [30].

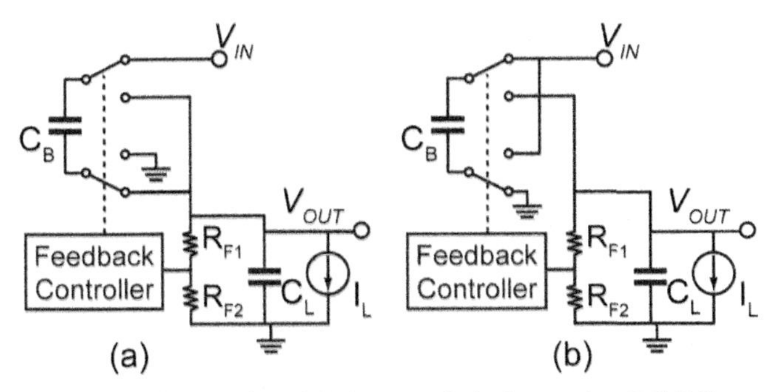

Figure 6.8 Schematic of (a) buck and (b) boost switched-capacitor DC-DC converters [31].

Figure 6.8 shows a switched-capacitor DC-DC converter topology, where electrical charge is transferred through capacitor C_B. A clock input controls the switch timing (i.e., open or closed). A feedback controller can be employed to control the clock duty cycle or frequency that drive the switches, and thus changes the energy transfer rate and regulates the output voltage. A switched-capacitor DC-DC converter does not require a bulky, external, low-ESR inductor. It therefore produces lower electromagnetic emissions and generates less ripple as there is no inductive switching noise [33]. It can operate in either step-up/down configuration on the fly. However, ripple is still present because charge is injected into the output node to maintain the desired voltage at each clock cycle. The conversion factor, or DC gain of the switched-capacitor DC converter is always a discrete number, because voltages across the capacitor are either added or subtracted in some quantized amount. The disadvantages of the switched-capacitor scheme are: it is less efficient ($<80\%$), requires more switches (i.e., higher circuit complexity), and efficiency depends upon the output voltage magnitude.

There is energy loss inherent in the charge transfer mechanism in a switched-capacitor circuit, which can be illustrated by examining the charge transfer process between two capacitors ($C_1$and C_2 in Figure 6.9). Initially, C_1 is charged to voltage V_1 and C_2 is charged to V_2 (see Figure 6.9(a)). The total energy stored in both capacitors is given by

$$E_a = \frac{1}{2}(C_1V_1^2 + C_2V_2^2). \tag{6.2}$$

When the switch is closed, charge is transferred from higher to lower potentials, creating a new equilibrium voltage, V_3 across both capacitors (see Figure 6.9(b)). V_3 can be calculated as

$$V_3 = \frac{C_1V_1 + C_2V_2}{C_1 + C_2}, \tag{6.3}$$

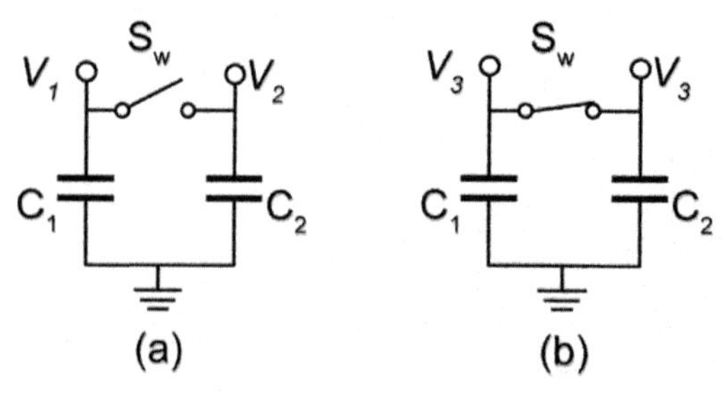

Figure 6.9 Illustration of charge transfers between two capacitors through a switch.

and the total energy stored in both capacitors is now given by

$$E_b = E_a - \frac{1}{2}\frac{C_1 C_2}{C_1 + C_2}(V_1 - V_2)^2. \tag{6.4}$$

As long as there is charge transfer in a switched-capacitor circuit, there is energy loss. Energy is lost due to dissipation in the finite resistance of the switch that depends on the initial voltage difference $V_1 - V_2$, where a bigger voltage difference results in higher loss. This loss limits the maximum efficiency that can be achieved by a switched-capacitor DC converter.

There are other sources of energy loss in a switched-capacitor DC converter, such as overhead in the peripheral circuitry, biasing, clock generation, and buffering [34]. The overhead should be much smaller than the power delivered to the load. Another source of energy loss is parasitic capacitances associated with the switches, e.g., C_{GS} and C_{GD} in a MOSFET switch. The gates of the switches must be charged and discharged at the clock rate, and the energy required to do that contributes to the loss. Parasitic capacitance at the bottom plate of the flying capacitor (e.g., C_B in Figure 6.8) is charged and discharged without contributing to charge transfer, and it also contributes to the total losses. Therefore, a high density capacitor with low parasitic capacitance between its top and bottom plates and the substrates is desirable for a switched-capacitor DC-DC converter.

The work in this Chapter will focus on a switched-capacitor regulator design, because a wide input range DC-DC converter is needed to convert the voltage across the supercapacitor to a supply voltage of 1 V. A stand-alone LDO is not suitable for the wide input voltage range anticipated as its output voltage is limited, and efficiency is (on average) less than that of a switching regulator. However, an LDO is still needed to supply sensitive analog and RF circuits that cannot tolerate ripple on their supply voltage. On the other hand, the switched-capacitor DC converter works across wide input voltage ranges and offers better average efficiency than the LDO, but it generates output ripple (typically tens of mV). A hybrid between the switched capacitor DC converter and LDO regulators is proposed to improve the overall efficiency with negligible ripple (below 0.1 mV) in the output voltage.

6.3 Power Management Circuit Design

The proposed power management testchip shown in Figure 6.11 was designed to harvest solar energy, store it on supercapacitor C_{SCAP}, and provide the supply voltages required by the wireless transceiver. A 225-mAh Li-ion (CR2032) battery at 3-V nominal voltage and feature size of 1 cm^3 is used as an auxiliary energy source. It provides energy during the start-up of the power management unit, and when energy is needed urgently, but is not available from the solant. A $2 \times 2 - \text{cm}^2$ amorphous silicon solant is employed to harvest the solar energy and act as an RF transducer [35]. The solant has a V_{OC} of 0.9 V and I_{SC} of 43 mA while radiated at the AM1.5 standard of 1000 W/m^2 and temperature of 25 °C [35]. The solant achieves P_{MAX} of 20 mW and 6.3% efficiency. Indoors, and near a window, the average power generated is 120 μW with an average V_{PMAX} of 0.42 V. Figure 6.10 shows power generated daily by the solant in a typical office during first week of November in Delft, the Netherlands. The solant is facing upward, near a window that faces the south-west direction. The energy generated daily varies by a factor of eight between cloudy and sunny days. The summer season on average generates twice the amount of energy compared to the winter season [36]. Note that the power generated daily due to office lighting (at 435 lux) is approximately 60 μW.

The switched-capacitor charge pump (SC-CP) (see Figure 6.11) transfers the charge generated by the solar cell to a supercapacitor. The supercapacitor size is chosen from an estimate of the energy that can be harvested from the solar antenna. Assuming indoor operation at an average generated power of 0.12 mW, the total energy generated daily by the solant is 10.4 Joule. From Equation (6.1), it is estimated that a 2.7 F supercapacitor with a voltage rating of 2.75 V could store the maximum energy collected per day. If we consider that part of the energy is consumed by the load, power harvester, and other monitoring circuitry, then a fraction of the maximum capacitor size is adequate. The HS130F 2.4-F supercapacitor used in this work has a leakage current of less than 2 μA, and is $3.9 \times 1.7\,\text{cm}^2$ in size (Cap-XX) [28]. The supercapacitor

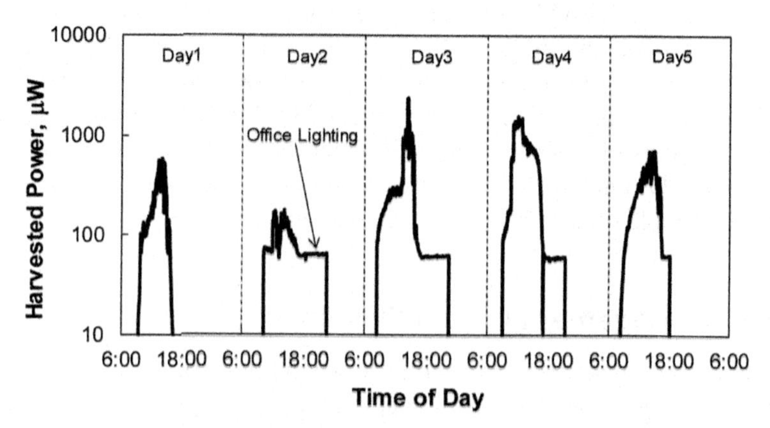

Figure 6.10 Power generated daily by the solar antenna (facing upward near a window that faces southwest) in an office at Delft, the Netherlands, during the 1st week of November 2011.

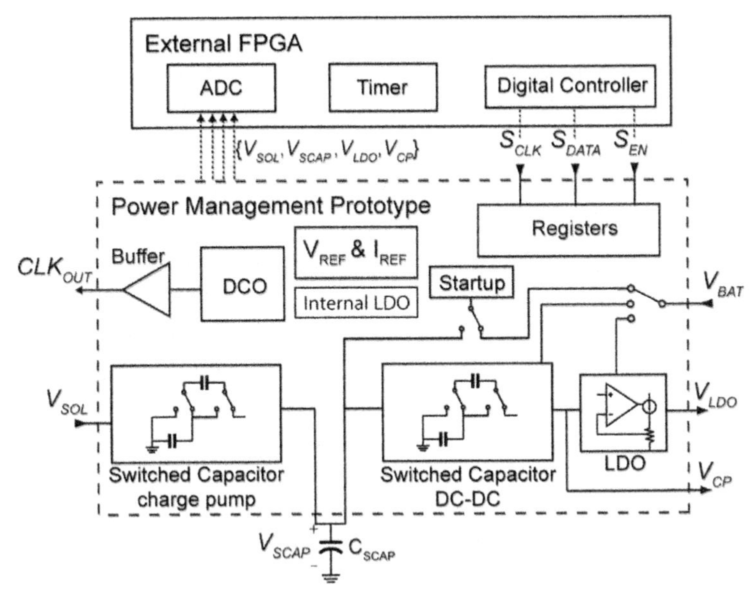

Figure 6.11 Block diagram of the power management test chip.

can be stacked with the solant, battery and the chip packaging, realizing a sensor node with an approximate volume of 5 cm^3.

The stored charge provides energy to a DC-to-DC converter that generates the required supply voltages. Three modes of DC-to-DC conversion are available. The first uses a low-dropout regulator (LDO), the second uses a switched-capacitor DC-to-DC converter (SC-DC), and the third uses a hybrid of the LDO and SC-DC converter, with performance intermediate to the other two supplies.

A digitally-controlled oscillator generates non-overlapping clocks required for the SC-CP and the SC-DC converter. Voltage and current references are also implemented on-chip to generate temperature insensitive references, and for general biasing. Serial registers store the digital data used to control all of the various blocks. Internal LDOs supply circuitry internal to the power management IC that operate below the battery voltage (i.e., 0.6 V and 1 V). The battery supplies these blocks initially through a startup network. After startup, the supply is then switched to V_{SCAP} when the supercapacitor is full. Digital blocks that are switching periodically in normal operation are operated at 0.6 V to minimize their power consumption. A complete implementation of a power management sub-system would employ a microcontroller, timer, and ADC functions in addition to the basic energy-management functions implemented in this prototype. In this work, an FPGA with an embedded ADC monitors the various input and output voltages, and controls the operation of the power management test chip (i.e., microcontroller function).

6.3.1 Switched-Capacitor Charge Pump (SC-CP)

The charge generated by the solar cell at its output V_{SOL} (see Figure 6.3) is transferred to a supercapacitor via a switched-capacitor charge pump (SC-CP). Here, the serial-parallel charge pump topology is adopted [37], other charge pump topologies have also been proposed, e.g., the Dickson charge pump [38] or the Cockcroft-Walton charge pump [39]. The charge pump operates in two phases and is driven by non-overlapping clocks. As shown in Figure 6.12, the solar cell charges capacitors $3C_{PUMP}$ in parallel during phase ϕ_1. During phase ϕ_2, capacitors $3C_{PUMP}$ are connected in series with V_{SOL}, and charge is transferred from $3C_{PUMP}$ to the supercapacitor C_{SCAP}. The charge pump of Figure 6.12 effectively multiplies the DC voltage V_{SOL} by two to output V_{SCAP} in the steady-state condition. The charge pump can be configured with a multiplication/gain of one when the output voltage is less than V_{SOL} by directly connecting V_{SOL} to V_{SCAP} through the switch controlled by V_{SH}. A DC gain of four is realized by the configuration shown in Figure 6.13, where each of the three capacitors C_{PUMP} is connected in series with V_{SOL} during the second phase. If the maximum value expected for V_{SOL} is 0.75 V, a maximum output voltage (V_{SCAP}) of 2.75V can be realized theoretically (limited by the supercapacitor voltage rating [28]).

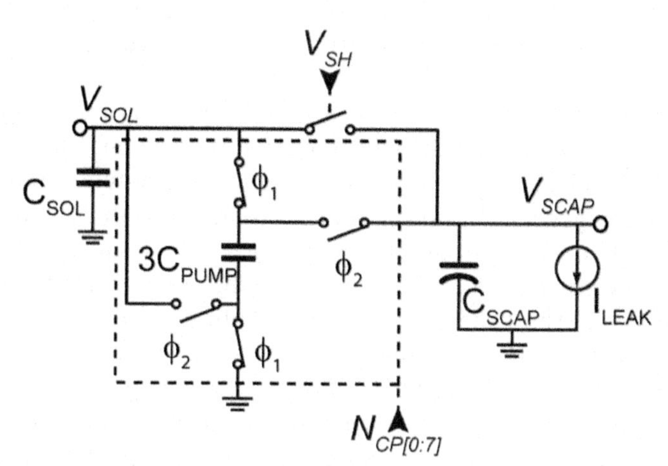

Figure 6.12 Switched capacitor charge pump (SC-CP) schematic.

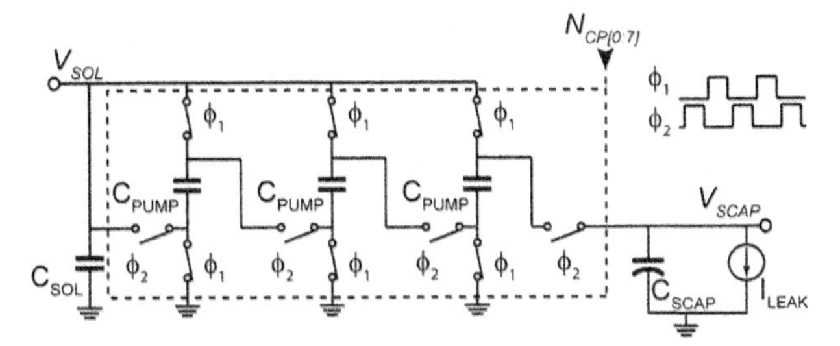

Figure 6.13 SC-CP with a DC gain configuration of four.

It is advantageous for the charge pump to load the solar cell so that the maximum available power can be obtained from the solar cell. The loading of the solar cell (R_{SOL}) by the switched capacitor circuit is given by

$$R_{SOL} = \frac{1}{3C_{PUMP}f_{clk}}\frac{1}{N_{CP}}, \tag{6.5}$$

where $3C_{PUMP}$ is the total pump capacitance used in the SC-CP (see Figure 6.12), and f_{CLK} is the clock rate of the charge pump, which can be varied to change R_{SOL}. In the power management testchip, there are seven parallel charge pumps controlled by a 3-bit digital control word, N_{CP}. Control word N_{CP} can also be varied to adjust the load on the solar cell (i.e., changing effective size of C_{PUMP}) to obtain optimum power transfer. The optimum load for the solar cell varies with the illumination and lies within the range between 0.1–10 kΩ (estimated from solant characterization). Figure 6.14 shows the measured power generated by the solant versus load impedance for varying light intensity. The optimum impedance that extracts maximum power is lower

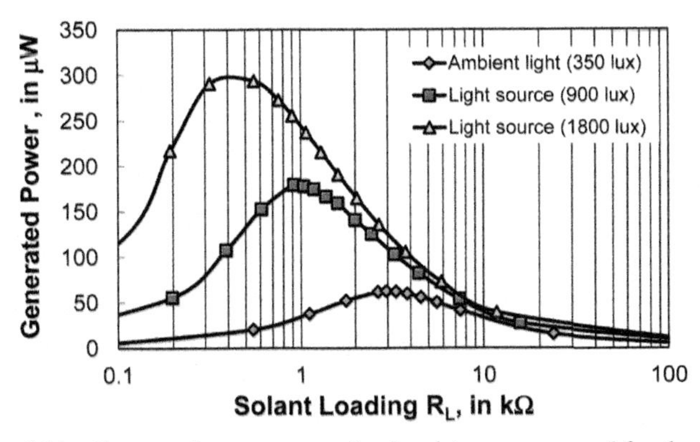

Figure 6.14 Generated power versus load resistance measured for the solant.

for a higher light intensity. The optimum (extrapolated) impedance for light indoors that comes from the sun via a window (measured intensity of around 10,000 lux) is 0.1 kΩ. The V_{PMAX} of a solar cell is estimated at approximately 80% of the V_{OC} [40]. Optimum energy transfer can be achieved by measuring its initial V_{OC} of the solar cell and incrementally reducing the input impedance of the power harvester until the solar cell output reaches V_{PMAX}. Based on the curve shown in Figure 6.14, the optimum voltage for maximum available power is around 0.4 V.

When the SC-CP is used in conjunction with a solar cell to harvest energy, the output voltage V_{SCAP} (i.e., voltage stored on the supercapacitor) will be in the range of 0-2.75 V. Due to its inherent loss, maximum efficiency is attained by a charge pump when the output voltage reaches the maximum value [41]. Therefore, efficiency is low until V_{SCAP} reaches four times V_{SOL} if a DC gain of four is employed. A multi-step mode using incremental gain is employed to ensure that the charge transfer process is as energy efficient as possible. In the multi-step mode, the SC-CP uses the smallest gain initially. The gain is then increased incrementally when the output reaches the maximum achievable voltage for each respective gain setting. Figure 6.15 compares the energy transfer efficiency of single-step and multi-step charge pumps when simulated with C_{SCAP} value 10 times C_{PUMP}. Only energy loss as defined by Equation (6.4) is considered in this simulation. The output charges exponentially, eventually reaching a final voltage of $4V_{SOL}$ in this case. Gain steps of 1, 2 and 4 are used in the multi-step example. Charging in multiple steps is slightly slower than when a single step is used, as shown in Figure 6.15(a). The time required to charge C_{SCAP} to its maximum voltage can be estimated from the 1/RC time constant at

$$T_{CH} = \frac{1}{RC} = \frac{C_{SCAP}}{C_{PUMP} f_{clk}}. \tag{6.6}$$

The energy loss per clock cycle is highest initially, and diminishes as the output charges (see Figure 6.15(b)). Figure 6.15(c) shows the cumulative energy loss as a percentage of the total energy stored on the output capacitor. While a single step charge pump loses 48% of the total transferred charge in the pumping process, energy loss is reduced to less than 20% by using three steps. A larger number of steps could reduce the inherent energy loss (asymptotically) to zero, in theory. However, it is not practical to realize this in practice, because it requires a large number of switches and capacitor configurations to generate the various gains. A more complex circuit suffers from increased parasitics, which reduces the overall power efficiency.

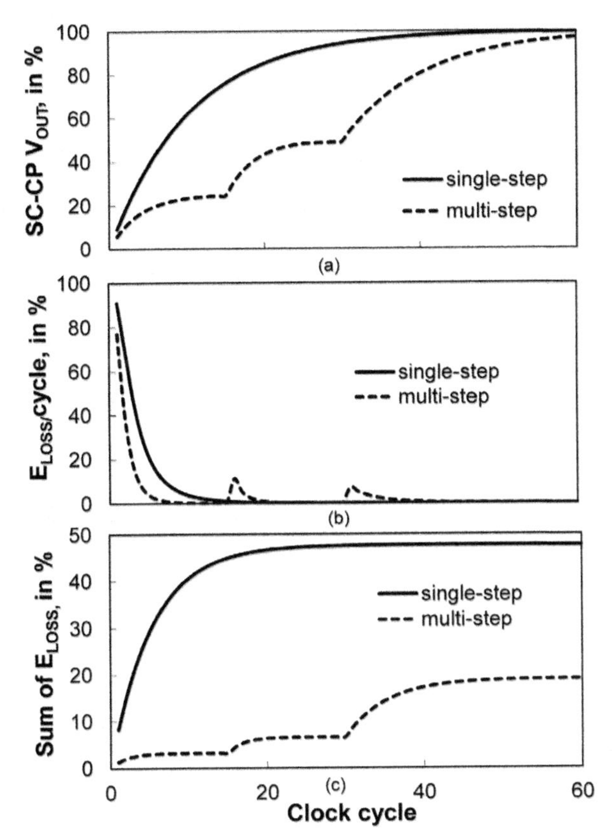

Figure 6.15 Simulated comparison of energy transfer loss for single-step, and multi-step charge pump circuits. (a) Output voltage, (b) Energy loss per cycle, and (c) Cumulative energy loss versus clock cycle.

An ADC samples the charge pump output voltage and an algorithm determines the best configuration for the charge pump gain, f_{CLK}, and N_{CP} (i.e., close to optimum within the limited number of configurations). The output voltage has to be monitored periodically, as the solar cell output varies due to changes in ambient lighting. If the charge pump operates at a smaller gain, charge could travel backward, discharging the supercapacitor instead of charging it.

An external 1-nF capacitor (C_{SOL}) is connected in parallel with the solar cell to reduce voltage fluctuations caused by switching of the switched-capacitor circuit. Each C_{PUMP} unit is implemented using an 18.6-pF MIM capacitor with 2% parasitic capacitance on its bottom plate. The nominal clock frequency (f_{CLK}) for the charge pump is 10 MHz. The power consumed by the charge pump is 15 μW and 105 μW for N_{CP} of 1 and 7, respectively.

6.3.2 Low Drop-Out (LDO) Regulator

The LDO design shown in Figure 6.16 supplies the load with 1 mA and 1 V under typical operating conditions. The output stage uses a zero-threshold NMOS source follower instead of a PMOS pass transistor as used in a conventional LDO [42]. A zero-threshold device (i.e., $V_T \approx 0$ V) allows low offset between the input and output, resulting in a dropout voltage of just 0.15 V. The source-follower configuration offers a good power supply rejection ratio (PSRR) at all frequencies as long the NMOS transistor is operating in the saturation regime [43]. The 3.6-pF MOS capacitor, M_{CAP}, stabilizes the loop, which has a phase margin of 90° for a capacitive load of 1 nF. The main regulating amplifier, A_1, uses a folded-cascode topology and has a DC gain of 52 dB and gain bandwidth product of 10 kHz as shown in Figure 6.17. Feedback resistors R_{F1} and R_{F2} are implemented externally using 1 MΩ potentiometers.

The extra circuitry shown in the dashed box in Figure 6.16 compensates for any fast transients (e.g., spikes) in the output voltage caused by load switching. For example, the load current in a duty-cycled radio has a pulse like profile which peaks during the active time. Because of the-low frequency (dominant) pole seen at V_{GATE} in Figure 6.16, the loop takes a few tens of microseconds

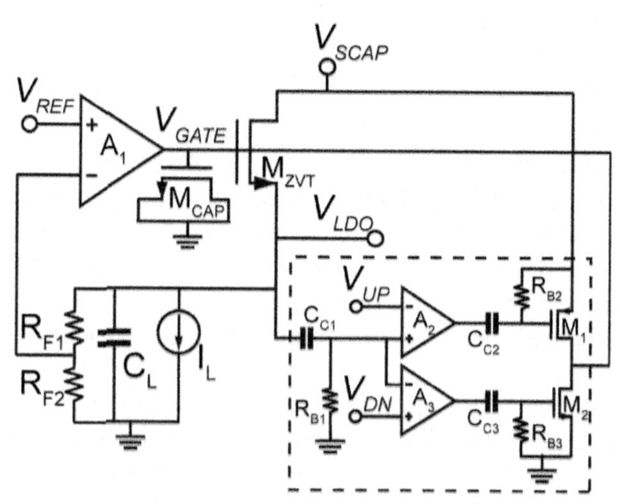

Figure 6.16 Schematic of the proposed LDO.

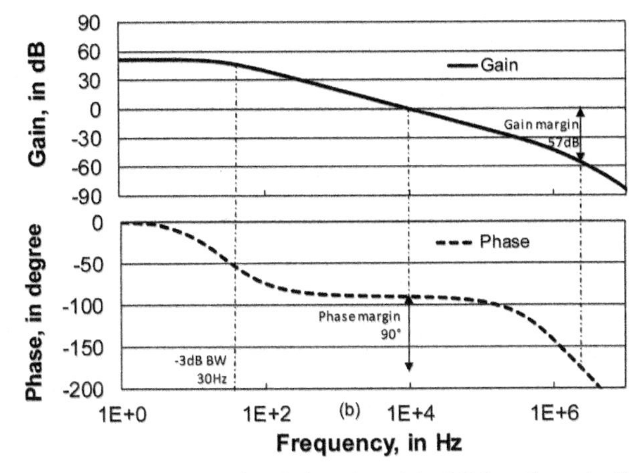

Figure 6.17 (a) Gain and (b) phase simulation plot of the LDO at $C_L = 1$ nF and $I_L = 1$ mA.

to respond when a large step in the load current occurs. Transient spikes are detected via an RC highpass network (C_{C1} and R_{B1}) and window comparators, A_2 and A_3. The pulse generated by this comparison pulls or pushes the voltage at node V_{GATE}, effectively speeding-up the loop response time. The amount of compensation is controlled by a 3-bit DAC (not shown) that tunes the speed of the comparators by varying its current bias. The LDO quiescent current (excluding current through $R_{F1\text{-}2}$) is 0.4 μA from a 3-V battery or V_{SCAP}. The usable voltage range of V_{SCAP} is 1.15-2.75 V.

6.3.3 Switched-Capacitor DC Converter (SC-DC)

The switched-capacitor DC-DC converter circuit uses a series-parallel topology [37]. The converter configuration shown in Figure 6.18(a) realizes a DC

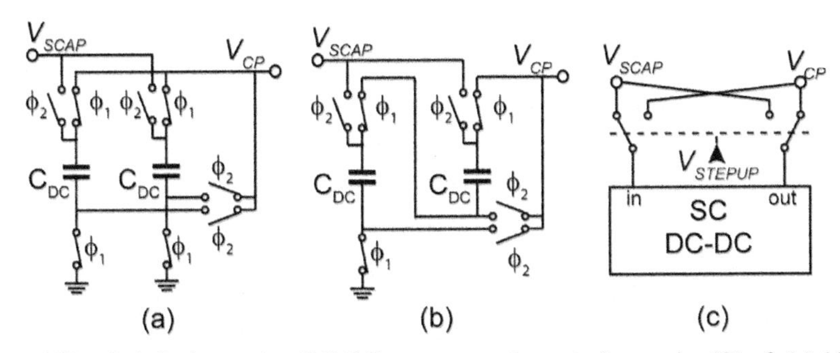

Figure 6.18 Switched-capacitor DC-DC converter schematic for a gain, (G) of: (a) 1/2, (b) 2/3, and (c) in the step-up configuration.

gain of one-half when operating as a step-down DC converter. In this case, the input is the voltage across supercapacitor, V_{SCAP}, and the output is its scaled voltage, V_{CP}. By switching capacitors C_{DC} between V_{CP} and ground in one phase, and between V_{SCAP} and V_{CP} during the other phase, V_{CP} is charged to one-half of V_{SCAP} in the steady state. This network can be reconfigured by changing the switch configuration to realize a gain of two-thirds, as shown in Figure 6.18(b). The step-up converter of Figure 6.18(c) is implemented by exchanging the input and output ports of the networks shown in Figure 6.18(a) and Figure 6.18(b), which realizes a gain of two, and a gain of one and a half, respectively. The gain (G) of the DC converter can be changed easily by changing the switch configuration, showing the versatility of the switched-capacitor DC-DC converter circuit. Variable gain (G) is needed to accommodate a range of possible input voltages, in this case the voltage across the supercapacitor, V_{SCAP}, which can range from 0 to 2.5 V.

To maximize efficiency, there are several aspects of the design that can be optimized. Choosing a capacitor with low bottom or top plate parasitics (e.g., MIM-cap) is beneficial. Intrinsic losses can be minimized by ensuring that the voltage drop across C_{DC} is small during loading by using a bigger capacitance value and/or a faster clock speed. However, a bigger capacitor requires larger switches in the converter, which actually increases losses in the switch drivers. Similarly, increasing the clock speed also increases losses in the driver circuit, due to charging and discharging of parasitic capacitances in the switch and bottom-plate capacitances. Extensive simulations were performed to balance intrinsic and switching losses. Depending on the magnitude of the parasitic capacitance, there is an optimum clock speed that results in maximum efficiency for a given capacitor size. The power consumption of the peripheral circuitry is minimized by using high-speed, low-voltage CMOS switches (to minimize transient currents), a low-voltage supply, and level shifting to a higher voltage only when necessary. From simulation, the intrinsic loss is estimated to be 10% and other losses add another 10%.

A wireless circuit requires a clean supply voltage free of ripple and harmonic energy that could corrupt the transmit/received data, or jam the receiver. A hybrid DC-DC converter that combines an SC-DC converter and an LDO regulator is proposed for power management of the FM-UWB transceiver. Efficiency is maintained above 50% over a wide input voltage range for the hybrid DC-DC converter compared to an LDO alone (see Figure 6.19). The measured efficiency of the SC-DC is 79%, and the LDO efficiency given by the ratio of the 1-V output and V_{CP} (LDO quiescent current is negligible). The usable input voltage range is extended by using a step-up

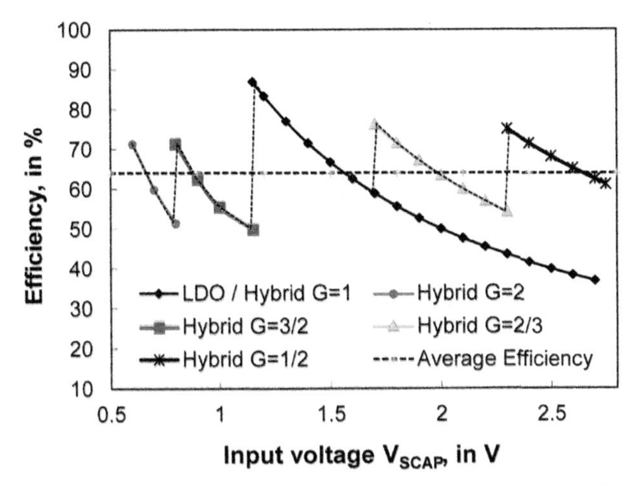

Figure 6.19 Efficiency of a hybrid DC-DC converter compared to an LDO as a function of the input voltage.

SC-DC converter configuration. The gain associated with each input voltage range is shown in Figure 6.19. The hybrid solution has an average efficiency of 64% for input voltages ranging from 0.55–2.75 V.

Figure 6.20 shows a block diagram of the hybrid DC-DC converter designed in this work for the autonomous wireless system. There is no feedback mechanism that regulates the SC-DC output, instead an LDO provides the voltage regulation. An ADC could be employed to monitor V_{SCAP} and V_{CP} when current loading is applied. A digital control is deployed to set the gain of the SC-DC, the clock frequency, and the N_{DC} such that the converter operates with optimal performance.

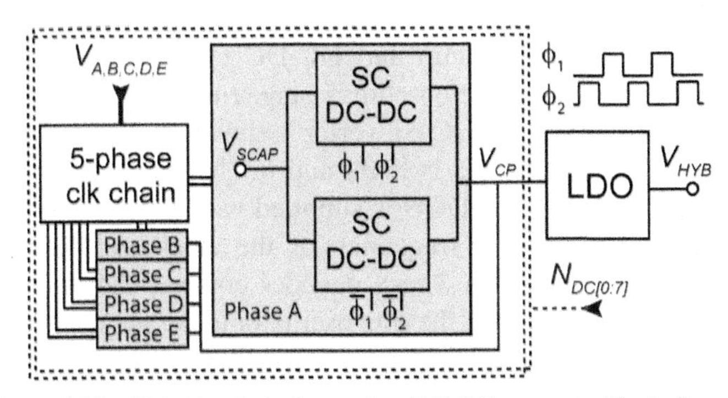

Figure 6.20 Hybrid switched-capacitor DC-DC converter block diagram.

Ripple is inherent in the SC-DC converter output, and can be as high as 30 mV$_{\text{p-p}}$ [44]. While an LDO may offer up to 40 dB of ripple rejection, this is not sufficient for the wireless receiver. Ripple should be less than 10 μV (equivalent to -90 dBm in a 50-Ω system), which implies 70-dB ripple rejection for the receiver's supply. Ripple is generated when charge is injected into the supply line to maintain the desired supply voltage at each clock cycle. Splitting the charge pump into a differential circuit, where each half-circuit works in tandem during opposite phases, reduces the ripple voltage at the output by a factor of 2, while maintaining the same time to charge the load. To further reduce the ripple voltage generated by the SC-DC converter, a multiphase clock is employed as shown in Figure 6.20. The 5-phase clock implemented in this prototype drives the SC-DC in time interleaving fashion. The multiphase clock not only reduces the magnitude of the ripple, but also pushes the ripple to a higher frequency where capacitance in the supply line can further filter the ripple. Reducing the ripple in the output voltage benefits efficiency as it also reduces ripple in the load current [45].

The efficiency typically reaches its maximum value at the designed power rating, and it is poorer at lower output powers. In this prototype, a unit DC-DC converter is designed that is optimized for output power of 0.25 mW. Unit converters could also be configured to operate in parallel, so that a wide range of load powers may be delivered efficiently. Scaling is controlled digitally by the 3-bit control signal N_{DC}, which is activated depending on the power demanded by the load, similar to method used in digital power amplifiers [46]. Each $\mathrm{C_{DC}}$ unit is implemented using MIM capacitors with a total capacitance of 25 pF. The nominal clock frequency (f_{CLK}) for the charge pump is 10 MHz. The power consumed by the SC-DC per unit ($\mathrm{N_{DC}} = 1$) is 21 μW.

6.3.4 Clock Generator

The switched-capacitor charge pump and DC-DC converter require non-overlapping, two-phase clocks. The on-chip, 5-stage ring DCO that generates the 5-phase clock used by the SC-DC converter is shown in Figure 6.21. The ring oscillator starts-up when V_{EN} is high, and the buffered clock outputs are $V_A, V_B, V_C, V_D,$ and V_E. The DCO is supplied with a regulated voltage of 0.5–0.6 V, which controls the frequency of the clock. Additionally, a 3-bit control signal fed through a 3-to-8 decoder controls the stage delay by adding inverters to each stage of the ring oscillator under the control of tri-state inverters. The frequency range of the DCO is 1–40 MHz and it consumes 1.4 μW when running at the highest frequency.

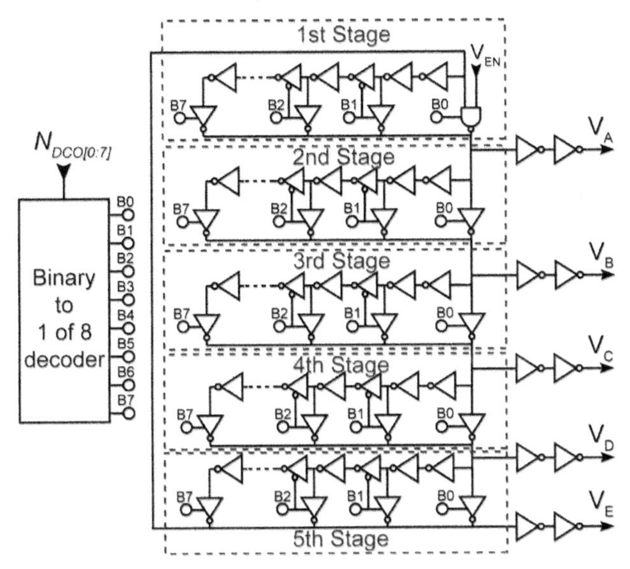

Figure 6.21 Schematic of the digitally-controlled oscillator (DCO).

The clock needs to be distributed to the switched-capacitor circuits. Figure 6.22 shows the clock tree from the DCO towards the switches. A multiplexer selects whether a 1 or 5-phase clock is used to drive the switched-capacitor circuits. A non-overlapping, two-phase clock is generated from each DCO output by a cross-coupled NAND flip-flop, as shown in Figure 6.23(a). Buffers and the non-overlapping generators are supplied from 0.6 V (typically) to take advantage of high-speed, low-threshold voltage transistors, and to minimize power consumption. The clock is shifted to a higher voltage (2.5–3 V) by the level shifter circuit shown in Figure 6.23(b). Resistor R_1 in Figure 6.23(b) speeds-up the switching process by bringing the common-mode voltage down dynamically during switching. Transistors M_{N1} and M_{N2} are low threshold devices to improve the circuit sensitivity to the 0.6-V

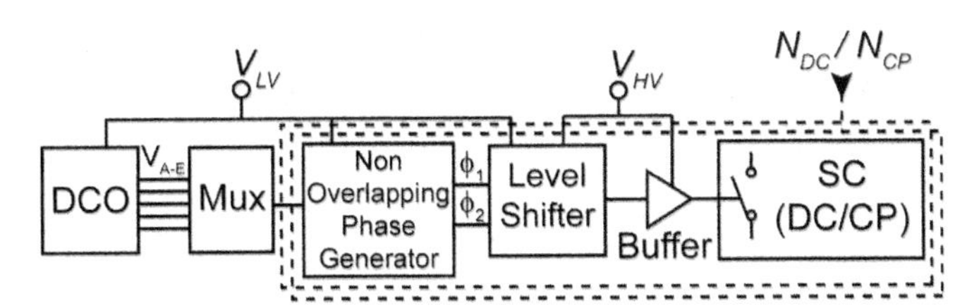

Figure 6.22 Clock distribution chain for the SC circuits.

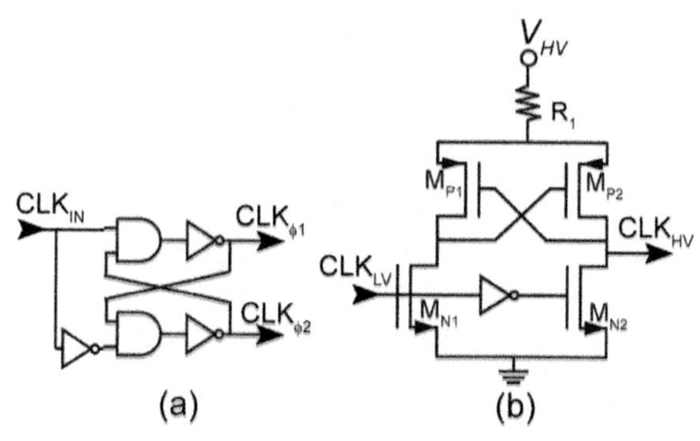

Figure 6.23 Schematic of (a) non-overlapping clock generator and (b) clock level shifter.

input voltage. A clock buffer drives the switches of the switched-capacitor charge pump, or the DC-DC converter. The central clock generated by the DCO is distributed to each unit of the switched-capacitor circuits, i.e., SC-DC and SC-CP. Each branch of SC-CP and SC-DC circuits (there are 7 branches requiring a 3-bit control word) has its own clock chain that is only turned on when needed to minimize the dynamic power consumption.

6.3.5 Bias Generator

Bias and voltage references are also generated on-chip. A schematic of the bias generator is shown in Figure 6.24. Transistor M_{N1} and $4M_{N1}$ loop creates

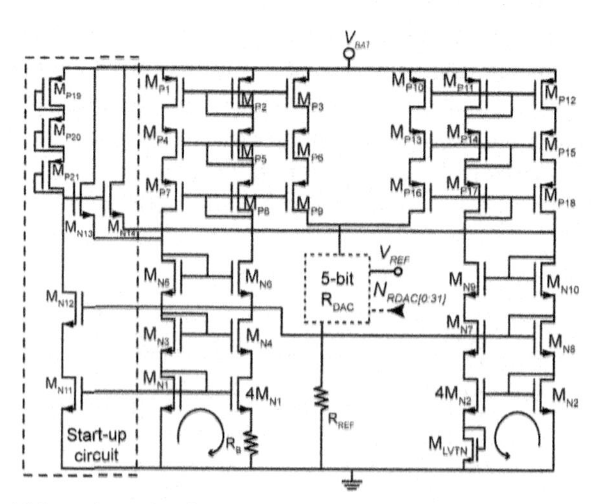

Figure 6.24 Schematic of the current bias and voltage reference generators.

a PTAT current, while the loop containing transistor M_{N2} and $4M_{N2}$ creates a current that is proportional to the threshold voltage of transistor M_{LVTN}, and hence is an inverse-PTAT current [47]. The sum of these currents produces a (first-order) temperature-independent current source, and bias currents are copied from this current reference. Since there is plenty of supply headroom, a triple cascode configuration is used to obtain 60-dB PSRR across a frequency range of DC to 100 kHz. A start-up circuit ensures that the loop operates at the desired operating point. A reference voltage for trimming purposes is created across R_{REF} from the current source using a 5-bit R_{DAC}. The reference voltage can be measured and calibrated to obtain a nominal value of 0.3 V. Each unit of the current bias is 50 nA, and it supplies the LDOs and current DACs. The bias generator consumes a total of 0.75 μA from a 3-V supply (battery, V_{BAT}).

6.4 Measurement Results

The power management testchip is implemented in a 90-nm bulk CMOS with the metal-insulator-metal (MIM) capacitor option [41]. MIM capacitors are used in the switched-capacitor DC-DC converter to store energy. The MIM capacitor is preferred to the back-end metal-oxide-metal (MOM) capacitor in this technology because it has higher density (3.1 fF/μm^2 versus 1.9 fF/μm^2), a lower parasitic capacitance (2% versus 15%), and a higher breakdown voltage (5.5 V versus 3.6 V). The all-copper interconnect scheme consists of 5 thin, two medium-thick, and a thick top metal. Thick-oxide MOS devices with a breakdown voltage of 3.3V are used for circuits that operate at the (nominal) battery voltage.

The 1.1 mm^2 (including bondpads) power management prototype is shown in Figure 6.25. Each pad is ESD protected using a standard library double-diode cell. The die is wirebonded in a 24-DIP and soldered onto a custom made PCB for testing, as shown in Figure 6.26. A potentiometer is employed as a feedback resistor to control the output voltage of the LDO. A switch is added to connect/disconnect the supercapacitor to the SC-DC. Figure 6.27(a) shows the PCB with an embedded solant that is used for various measurements and testing. More detailed characterization of the solant is reported in [35]. The power management PCB, Li-ion button cell battery, and the 2.8 F Cap-XX supercapacitor are stacked on top of each other as shown in Figure 6.27(b).

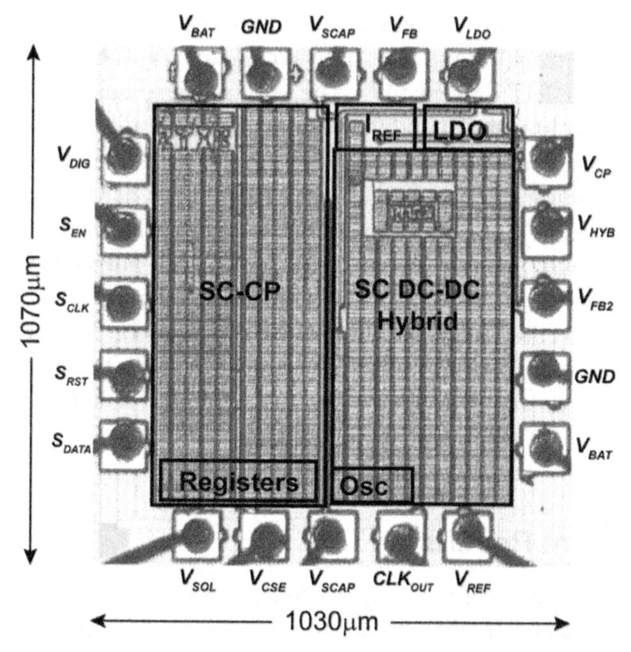

Figure 6.25 Chip photograph of the power management prototype.

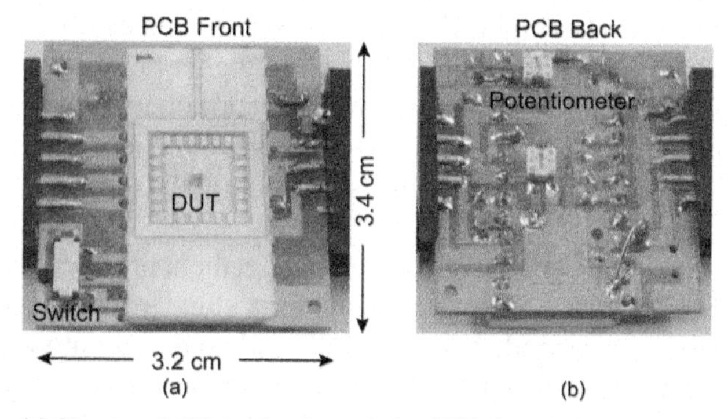

Figure 6.26 (a) Front and (b) back view of the PCB board for the power management characterization and testing.

6.4.1 SC-CP Characterization

The measured output power and efficiency of the SC-CP are shown in Figure 6.28 at f_{CLK} of 10 MHz, and DC-gain settings of 2 and 4 for an input voltage of 0.625 V. Power efficiency gradually improves because the loss inherent to the switched-capacitor circuit diminishes when the output voltage

Figure 6.27 Photograph of the (a) solant PCB and (b) combination of power management (PMAN) board with the Li-ion battery and supercapacitor.

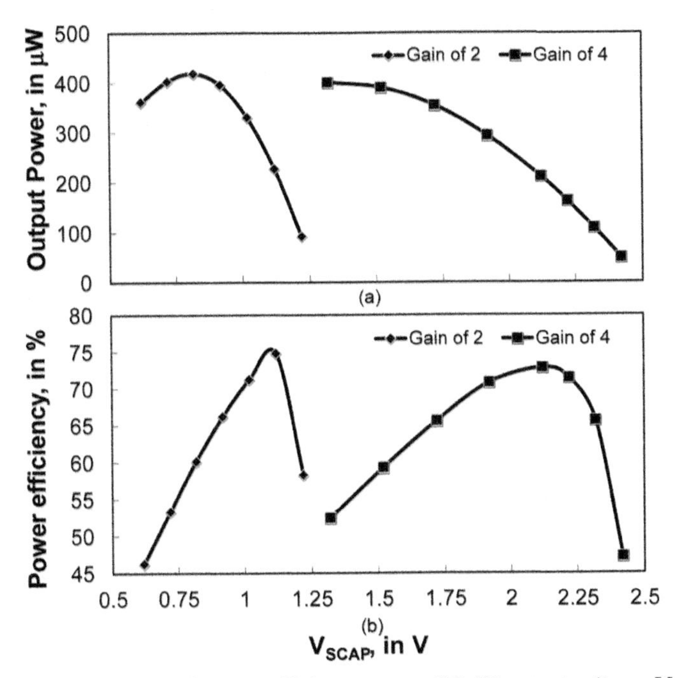

Figure 6.28 Measured power efficiency versus SC-CP output voltage, V_{CAP}.

is close to its maximum value. At this point, there is less power delivered to the load. At the same time, the clock is still running and consuming power, hence, efficiency peaks just before it reaches its maximum voltage output. The power efficiency reaches a peak at output voltages (V_{SCAP}) of 1.1 V and 2.15 V for DC gains of two and four, respectively.

Figure 6.29 shows the measured input resistance of the SC-CP at the minimum and maximum number of parallel charge pumps (N_{CP} of 1 and 7) for a clock frequency range of 6–18 MHz. The SC-CP has an input resistance range of 150-3,500 Ω, which fits the optimum load range for the solant under indoor

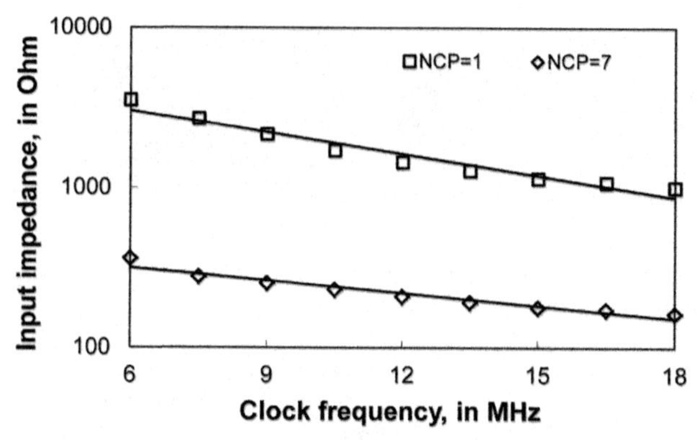

Figure 6.29 Measured input resistance of the SC-CP.

conditions (see Figure 6.14). Table 6.3 summarizes the performance of the SC-CP compared to other power harvester circuits. The SC-CP is customized for the solant, which generates a single diode voltage for input to the charge pump. The output range fits the supercapacitor voltage rating. The SC-CP achieves a peak power efficiency of 75%, which is lower than the power harvester reported in [49] because the clock frequency is above 5 MHz. A higher clock frequency means more power is dissipated in the clock generator, buffer, and parasitics. On the other hand, the SC-CP has an advantage of a small chip area of 0.32 mm^2 and a higher output power of 0.4 mW. There is a trade-off between clock frequency and area for a given power capacity.

6.4.2 LDO Characterization

Figure 6.30 shows the measured regulation of the LDO at an output voltage of 1 V for various load currents (I_L). The voltage regulation at 2.5 V is

Table 6.3 Performance summary of SC-CP

Parameters	This Work	[48]	[49]
Technology	90 nm CMOS	350 nm CMOS	350 nm CMOS
Input Voltage (V)	0.4–0.8	1–8	1–2.7
Output voltage (V)	0–2.75	1–8	2
Clock Frequency (MHz)	1–20	0.06	0.05
Max. Output Power (μW)	400	62.5	80
Peak efficiency (%)	75	58	86
Active area (mm^2)	0.32	3.06	2.28

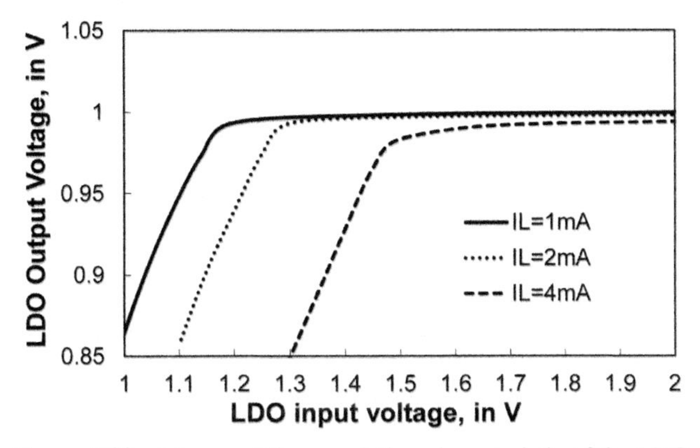

Figure 6.30 Measured line regulation characteristic of the LDO.

1.25 mV/mA. At a higher load current, the regulation diminishes and the dropout voltage increases. The dropout voltage is 0.15, 0.25, and 0.45 V for I_L of 1, 2, and 4 mA, respectively. Figure 6.31 shows measurement and simulation of the power supply rejection of the LDO versus frequency. At lower supply voltages, PSRR decreases because the NMOS pass device of the LDO operates closer to triode and hence the isolation from source-drain decreases. The poorer PSRR of 21 dB is measured at 50 kHz, and it is worse than simulation as it is probably caused by the lower output impedance of the NMOS pass device.

Figure 6.32 shows the measured step response of the LDO output voltage for a pulse current load of 2 mA. There is a spike in the output voltage when the current load changes suddenly due to the slow response of the opamp.

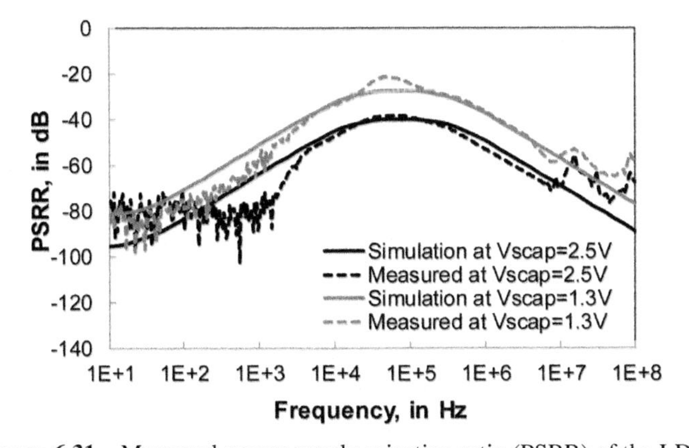

Figure 6.31 Measured power supply rejection ratio (PSRR) of the LDO.

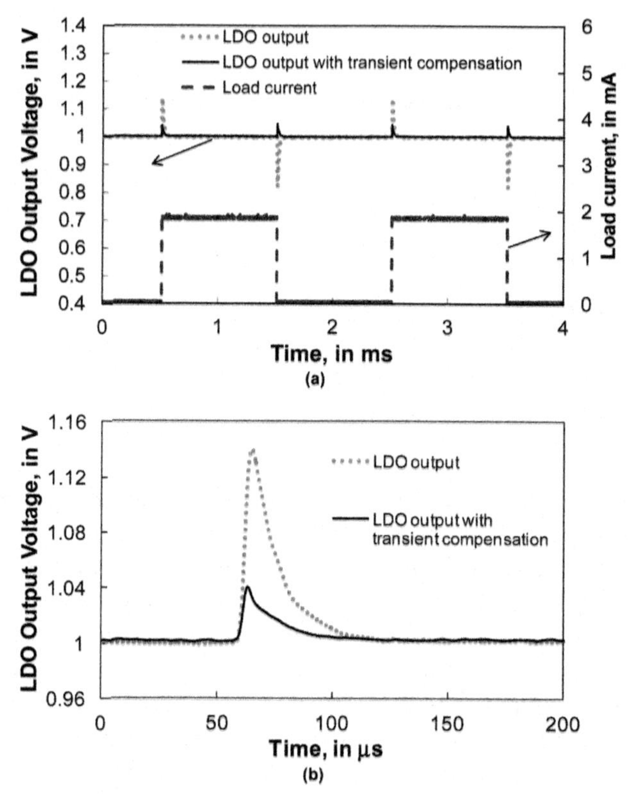

Figure 6.32 (a) Measured load transient response of the LDO, and (b) zoom-in on the voltage spike.

The output voltage variations due to the 2-mA current pulse are 0.14 V and −0.2 V at the rising and falling edges, respectively. The settling time is 60 μs. The aforementioned fast transient compensation circuit helps to reduce the spike to only 0.04 V for both rising and falling edges, and improves the settling time to 30 μs.

Table 6.4 lists all the important parameters of the proposed LDO in comparison to other published results from the literature. The proposed LDO is tailored to a small current load of 1 mA from a 1-V supply, which is the estimated current consumption of the FM-UWB transceiver. The LDO has a PSRR of 55 dB at 1 MHz, and fast settling of 30 μs, while consumes a quiescent current of only 400 nA. The LDO described in [51] has a smaller quiescent current, but at the cost of a much longer settling time. The proposed LDO has a low dropout voltage of 0.15 V, and the smallest chip area. It is

Table 6.4 Performance summary of the LDO regulator

Parameters	This Work	[50]	[51]	[52]
Technology	90 nm CMOS	130 nm CMOS	350 nm CMOS	350 nm CMOS
Active Area (mm^2)	0.013	0.049	0.9	0.264
Dropout Voltage (V)	0.15	0.15	0.2	0.2
Maximum Load (mA)	1	25	120	200
Quiescent Current (μA)	0.4	50	0.12	20
PSRR, at 1MHz (dB)	55	67	–	–
Load Regulation (mV/mA)	1.25	0.048	0.006	0.17
ΔV_{OUT} in transient (mV)	40	26	150	54
Settling time (μs)	30	–	1900	–

suitable to power a lightly loaded (<1 mA) circuit block that uses duty cycling rather than continuous operation.

6.4.3 SC-DC Characterization

Power efficiency of the SC-DC has been measured versus load current for different gain settings, number of units in parallel, and clock frequencies. Figure 6.33 shows the measured power efficiency at maximum N_{DC} and various gain settings. For the same I_L, the step-up converter requires more input current than the step-down converter. Therefore, peak efficiency of the step-up setting in the SC-DC occurs at a smaller I_L for the same capacitor size compared to the step-down setting. The measured peak power efficiency is 78.3%, 79.1%, 78% and 77.2% at gain settings of 3/2, 2, 1/2, and 2/3, respectively.

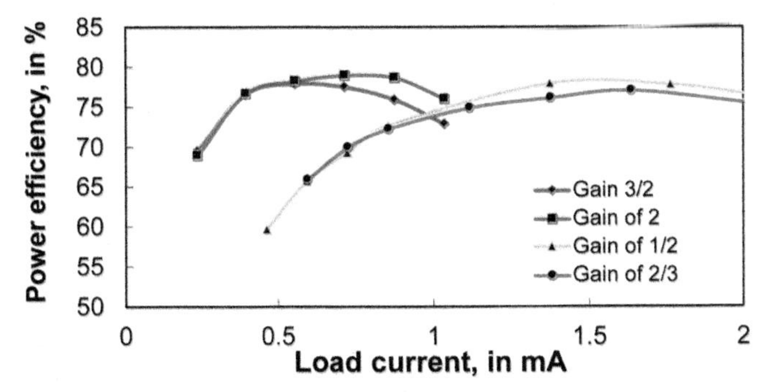

Figure 6.33 Measured power efficiency of the SC-DC for different gain settings at a clock frequency of 10 MHz.

Figure 6.34 shows power efficiency for different numbers of parallel SC-DC units as controlled by N_{DC}. The result shows that the peak efficiency can be controlled and maximized for the required current load. A power efficiency above 75% can be maintained over a wide range of load current. The range across which efficiency peaks can be increased further by changing the clock frequency. Power efficiency at a gain setting of 2/3 and maximum N_{DC} for various clock frequencies is plotted in Figure 6.35. A clock slower than the nominal of 10 MHz shifts the peak efficiency to a smaller load current, and vice versa. However, the highest peak efficiency was achieved at the nominal clock frequency.

As mentioned previously in Section 6.3.3, a multi-phase clock could reduce the ripple voltage generated by the SC-DC. Figure 6.36 shows the spectrum measured at the SC-DC output. The top figure shows the spectrum measured using a 1-phase clock signal. A peak tone of around 1 mV at 20 MHz appears together with other harmonics and sub-harmonics. When the SC-DC uses the 5-phase clock, the main tone due to ripple at 20 MHz is reduced to around

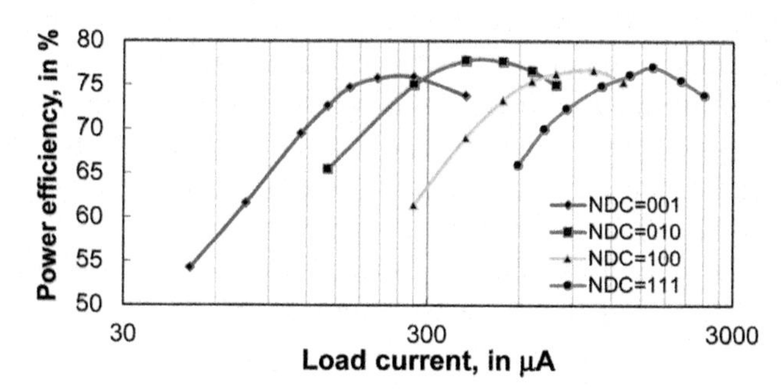

Figure 6.34 Measured power efficiency of the SC-DC for different numbers of parallel units (as controlled by N_{DC}) at a gain setting of 2/3 and clock frequency of 10 MHz.

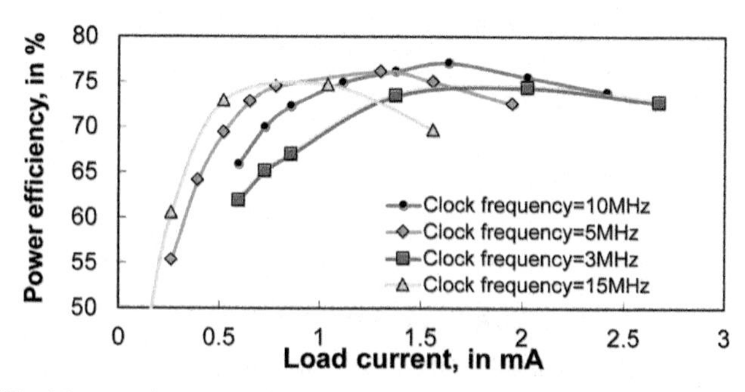

Figure 6.35 Measured power efficiency of the SC-DC for different clock frequencies.

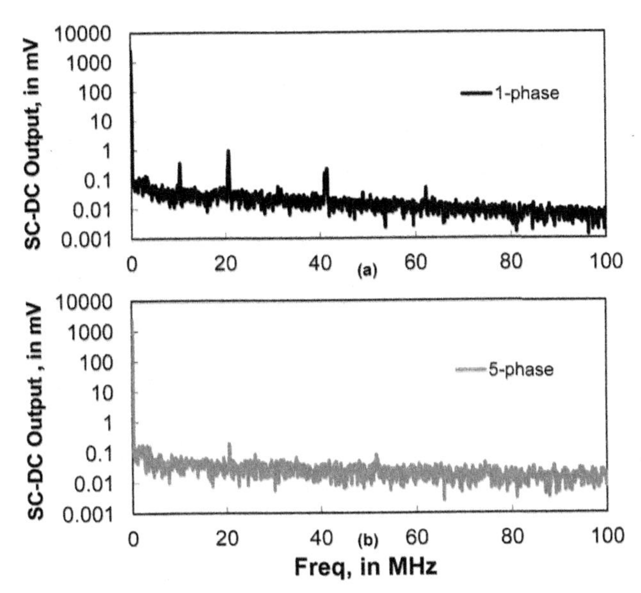

Figure 6.36 Measured spectrum on the SC-DC output voltage using (a) 1-phase and (b) 5-phase clocks at 10 MHz.

0.1 mV. The LDO, with a PSRR better than 40 dB, will further suppress this ripple to below 1 μV.

Figure 6.37 shows the transient performance of the hybrid SC-DC/LDO for a load current pulse of 1 mA. There is no regulation in the SC-DC, so the output voltage drops when the load current increases. However, the output

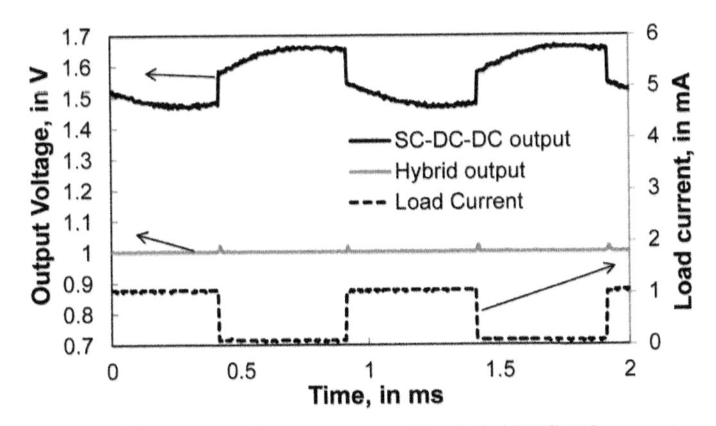

Figure 6.37 Measured step response of the hybrid DC-DC converter.

voltage of the hybrid generates 25 mV variations due to the 1 mA load current pulse which is regulated by the LDO.

Table 6.5 lists all the important parameters of the SC-DC converter in comparison with other published work. The SC-DC gets an input voltage from the charge stored on the supercapacitor in the range of 0.7–2.75 V. The output voltage for the FM-UWB transceiver load is 1 V. The SC-DC achieves 79.1% power efficiency, which is comparable to other SC DC-to-DC converters from the recent literature. Ripple voltage is suppressed using the on-chip 5–phase clock, generating only 0.1 mV, which is lower than the other converters. The power density is considerably better than other converters at 3.7 mW/mm^2. The power density is usually limited by the technology (i.e., availability of high-density, low-parasitic capacitors).

6.4.4 Power Management Sub-System

The power management sub-system combines all the building blocks described previously. The calculated voltage level across the 2.8 F supercapacitor (V_{SCAP}) is shown in Figure 6.38, based on the measured solant energy that was collected indoors (see Figure 6.10). Leakage current at V_{SCAP} is estimated based on the standby current consumed by the power management unit and the measured leakage of the supercapacitor (~5 µA). The SC-CP is assumed to charge the supercapacitor at an efficiency of 75%. In this case, the load current of 1 mA is applied when the V_{SCAP} above 0.7 V for one minute out of every 10 minutes (duty cycle of 10%).

A simplified flow chart of the algorithm used by the microcontroller to operate the power management unit is shown in Figure 6.39. The controller detects the available energy, stores it in the supercapacitor and supplies the load when there is enough stored energy. As expected, data throughput is

Table 6.5 Performance summary of the SC-DC converter

Parameters	This Work	[34]	[44]	[53]
Technology	90 nm CMOS	180 nm CMOS	130 nm CMOS	130 nm CMOS
Input Voltage	0.7–2.75	1.8	2.5–3.6	1.2
Output voltage (V)	1	0.3–1.1	0.44	0.7–2.1
Maximum Load (mW)	2	1	2.5×10^{-4}	4
Clock Frequency (MHz)	10	15	2×10^{-3}	30
Peak efficiency	79.1%	81%	56%	84.3%
Peak ripple (mV)	0.1	–	50	–
Active area (mm^2)	0.27	0.57	0.26	4
Power Density (mW/mm^2)	3.7	1.75	1×10^{-3}	1

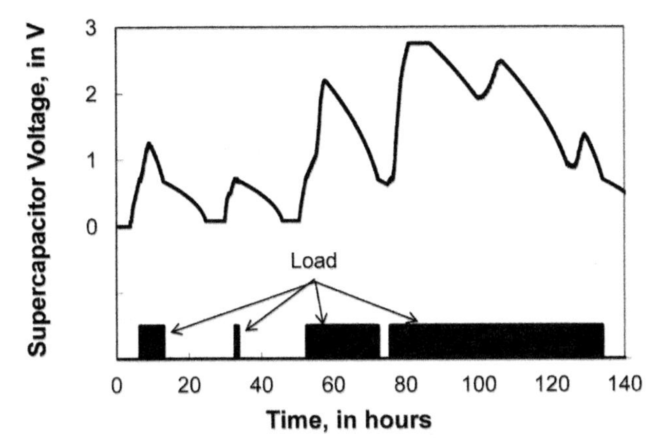

Figure 6.38 Calculated available voltage across the supercapacitor from indoor solar energy generated by the solant.

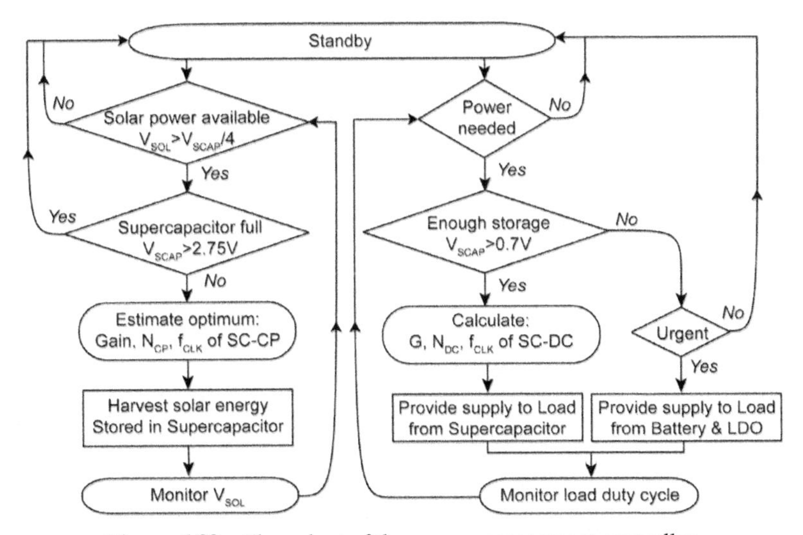

Figure 6.39 Flow chart of the power management controller.

intermittent when there is not much energy stored during startup. Later on, when the supercapacitor is almost full, transmission is more reliable.

The battery as a backup power source is essential for any urgent or emergency transmissions. The energy from the battery can also be used during startup. The lifetime of the battery depends on the leakage power (e.g., standby power in the power management, self-leakage of the battery) and how often the battery is used by the load. Figure 6.40 shows a 225-mA-hour battery lifetime

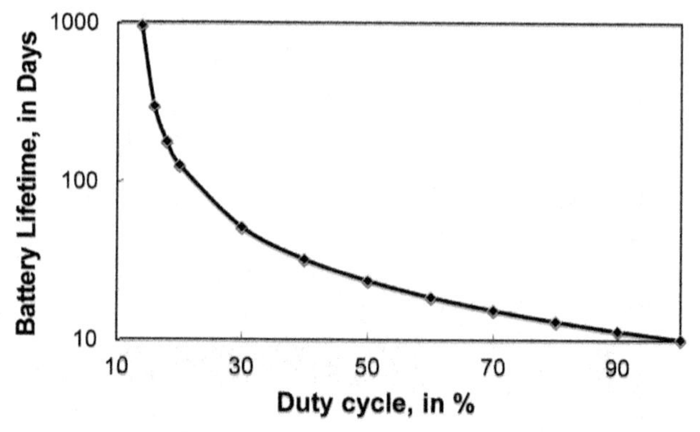

Figure 6.40 Estimated lifetime for a 225 mA-hour Li battery when using the collected solant energy and loaded by the FM-UWB transceiver (see Chapter 5).

versus duty cycle of the FM-UWB transceiver load described in Chapter 5. The average power collected from the solar cell is assumed to be 180 μW (averaging winter and summer condition), and the total leakage current is 10 μA. At 14% duty cycle, the battery lifetime is 3 years, while at 13% (i.e., when collected power equals consumed power) the lifetime is limited by the battery shelf life (~10 years). The solar energy that could be collected is on average 15.6 Joule per day, which is enough to supply a 6 nJ/bit wireless transceiver to transmit or receive raw data at 2.6 Gbit per day.

6.5 Conclusions

A power management prototype, suitable for autonomous wireless applications when powered from a button cell battery and solar antenna, has been realized in a low-cost, 90-nm bulk CMOS technology. The 1.1-mm^2 IC demonstrator consumes just 3.6 μA of stand-by current when supplied from a 3-V, 225-mA-hr button-cell battery allowing 7 years of standby time (theoretically). A 2.8-F Cap-XX supercapacitor is employed to store the energy generated by the solant. The switched-capacitor charge pump achieves 75% peak efficiency at a load resistance in the range of 0.15-4 kΩ. The hybrid DC-DC converter achieves an average efficiency of 64% from an input voltage range of 0.7–2.75 V, and output ripple is less than 1 μV. The battery lifetime (estimated), when loaded by the 0.6-mW transceiver described in Chapter 5, is 10 days when operated continuously, and 3 years at 14% duty cycle.

A future version of the power management sub-system would include an integrated, low-power ADC and an embedded micro-controller. A finite state machine that controls various building blocks in the sub-system will also be developed to optimize the power efficiency in real time based on environmental conditions. The power efficiency could be improved to 85% by using a slower clock speed at the cost of a larger die area [53]. The SC-DC converter is programmable and scalable. The SC-CP and the SC-DC could be implemented as multiple SC circuit cells that share the same capacitor arrays. The number of cells assigned for the CP and DC-DC could be partitioned based on harvested power availability and load demand. By merging SC-CP and SC-DC, the total chip area could be minimized.

References

[1] N. Henze, M. Waitz, P Hofmann, C. Bendel, K. Kirchhof, H. Fruchting, "Investigation on planar antenna with photovoltaic solar cells for mobile communications," *International symp. on Personal, Indoor and Mobile Radio Communications*, Vol. 1, Sept. 2004, pp. 622–626.

[2] S. Shynu, M. J. Roo Ons, M. J. Ammann, S. McCormack, B. Norton, "A metal plate solar cell for UMTS pico-cell base station," *Loughborough Antenna & Propagation Conference*, March 2008, pp. 373–376.

[3] N. Henze, A. Giere, H. Fruchting, P. Hofmann, "GPS patch antenna with photovoltaic solar cells for vehicular applications," *Vehicular Technology Conference*, Vol. 1, 2003, pp. 50–54.

[4] T. Wu, R. Li, M. M. Tentzeris, "A scalable solar antenna for autonomous integrated wireless sensor nodes," *IEEE antenna and wireless propagation letter*, Vol. 10, pp. 510–513, 2011.

[5] A. Valenzuela, "Batteryless energy harvesting for embedded design," *Texas Instrument design article*, Aug. 2009. Available: http://www.eetimes.com/design/embedded/4008326/Batteryless energy-harvesting-for-embedded-designs

[6] M. Danesh, J. R. Long, and M. Simeoni, "Small-area solar antenna for low-power UWB transceivers," *4th Eur. Conf. on Ant. and Prop. EuCAP'2010 Digest*, April 2010, pp. 1–4.

[7] J. Polastre, R. Szewczyk, D. Culler, "Telos: enabling ultra-low power wireless research," *International Symposium on Information Processing in Sensors Network*, April 2005, pp. 364–369.

[8] N. Saputra, J. R. Long, "A fully-integrated, short-range, low data rate FM-UWB transmitter in 90 nm CMOS," *IEEE Journal of Solid State Circuits*, Vol. 46, No. 7, pp. 1627–1635, July 2011.

[9] N. Saputra, J. R. Long, J. J. Pekarik, "A 2.2 mW regenerative FM-UWB receiver in 65 nm CMOS," *proc. IEEE RFIC Symposium*, May 2010, pp. 193–196.

[10] R. E. I. Schropp and M. Zeman, *Amorphous and microcrystalline solar cells: modeling, materials, and device technology*, Kluwer Academic Publishers, 1998.

[11] S. Beeby, N. White, *Energy harvesting for autonomous systems*, Artech House, 2010.

[12] Best Research cell efficiencies. Available: http://optics.org/news/1/5/5/recordcell

[13] R. E. I. Schropp, M. Zeman, "New developments in amorphous thin-film silicon solar cells," *IEEE Tran. on Electron Devices*, Vol. 10, pp. 2086–2092, Oct. 1999.

[14] N. H. Reich, W. van Sark, E. A. Alsema, et al., "Weak light performance and spectral response of different solar cell types," *Proc. of the 20th European Photovoltaic Solar Energy Conference and Exhibition*, 2005.

[15] R. C. Campbell, "A Circuit-based photovoltaic array model for power system studies," *North American Power Symposium*, Oct. 2007, pp. 97–101.

[16] T. Markvart. *Solar Electricity*. John Wiley and Sons, 1994.

[17] S. Buller, M. Thele, R. De Doncker, E. Karden, "Impedance based simulation models of supercapacitor and Li-Ion batteries for power electronic applications," *IEEE Trans. on Industry Applications*, Vol. 41, No. 3, May 2005, pp. 742–747.

[18] G. Martin, "Wireless sensor solution for home & building automation – the successful standard uses energy harvesting," *Enocean white paper*, Aug. 2007.

[19] F. M. Gonzalez-Longatt, "Circuit based battery models: a review," *Proceedings of 2nd Congreso IberoAmericano De Estudiantes de Ingenieria Electrica*, 2006.

[20] M. Pedram, Q. Wu, "Design consideration for battery-powered electronics," *Proc. Design automation conference*, June 1999, pp. 861–866.

[21] M. T. Penella, M. Gasulla, "Battery squeezing under low-power pulsed loads," *Proc. IEEE International Instrumentation and Measurement Technology Conference*, May 2008, pp. 1184–1188.

[22] S. Park, A. Savvides, M. B. Srivastava, "Battery capacity measurement and analysis using lithium coin cell battery," *International Symp. on Low Power Electronics and Design*, Aug. 2001, pp. 382–387.

[23] Battery life and how to improve it. Available: http://www.mpoweruk.com/life.htm

[24] Maxim application note, "Lithium coin-Cell batteries: predicting an application lifetime," March 2002.

[25] Seiko Instrument Micro Batteries Product Catalogue. Available: www.sii.co.jp/compo/catalog/battery_en.pdf

[26] Battery technical specifications. Available: http://www.microbattery.com/tech-specs.htm

[27] Energy storage technologies: a comparison. Available: http://www.cap-xx.com/resources/reviews/strge_cmprsn.htm

[28] Cap-XX H-series supercapacitors product bulletin. Available: http://www.cap-xx.com/resources/resources.htm

[29] Maxwell technology ultracapacitors. Available: http://www.maxwell.com/products/ultra capacitors/product.aspx?PID=PC10-SERIES

[30] D. M. Mitchell, *Dc-Dc switching regulator analysis*, New York: McGraw-Hill, 1988.

[31] A. Rao, W. McIntyre, J. Parry, U. Moon, G. Temes, "Buck-boost switched-capacitor DC-DC voltage regulator using delta-sigma control loop," *International Symp. on Circuit and System*, Aug. 2002, pp. 743–746.

[32] "Application note 1197 selecting inductors for buck converter." Available on: http://www.ti.com/lit/an/snva038a/snva038a.pdf

[33] M. Renaud, Y. Gagnon, "Inductorless versus inductor-based integrated switching regulators: bill of material, efficiency, noise and reliability comparisons," Integration Dolphin Inc. http://www.design-reuse.com/articles/19850/inductorless-inductor-based-integrated-switching-regulator.html.

[34] Y. K. Ramadass, A. P. Chandrakasan, "Voltage scalable switched capacitor DC-DC converter for ultra-low-power on-chip applications," *IEEE 2007 Power Electronics Specialists Conference*, June 2007, pp. 2353–2359.

[35] M. Danesh, J. R. Long, "An autonomous wireless sensor node using a solar cell antenna for solar energy harvesting," *Digest of IEEE Int. Microwave Symposium*, June 2011, pp. 1–4.

[36] Seasonal variation of the output of the solar panels at AT&T Park in San Francisco. Available: http://en.wikipedia.org/wiki/File:ATTParkannualoutput.png

[37] M. D. Seeman, S. R. Sanders, "Analysis and optimization of switched-capacitor DC-DC converters," *IEEE transactions on power electronics*, Vol. 23, No. 2, pp. 841–851, March 2008.

[38] J. F. Dickson, "On-chip high-voltage generation in MNOS integrated circuits using an improved voltage multiplier technique", *IEEE Journal of Solid State Circuits*, Vol. 11, No.3, pp. 374–378, June 1976.

[39] J. D. Cockcroft, E. T. Walton, "Production of high velocity positive ions," *Proceedings of the Royal Society*, Vol. 136, pp. 619–630, 1932.

[40] L. Brunneli, "Photovoltaics scavenging systems: modeling and optimization," *Microelectronics Journal*, Vol. 40, pp. 1337–1344, 2009.

[41] B. Jagannathan, R. Groves, D. Goren, *et al.*, "RF CMOS for microwave and mm-wave applications", *Proc. of Silicon Monolithic Integrated Circuits in RF Systems*, Jan. 2006, pp. 259–264.

[42] G. A. Rincon-Mora and P. A. Allen, "A low-voltage, low quiescent current, low drop-out regulator," *IEEE Journal of Solid-State Circuits*, Vol. 33, No. 1, pp. 36–44, Jan. 1998.

[43] W. Kruiskamp, R. Beumer, "Low drop-out voltage regulator with full on-chip capacitance for slot-based operation," *Proc. of ESSCIRC*, Sept. 2008, pp. 346–349.

[44] M. Wieckowski, G. K. Chen, M. Seok, D. Blaauw, D. Sylvester, "A hybrid DC-DC converter for sub-microwatt sub-1V implantable applications," *Digest. Symposium on VLSI*, 2009.

[45] H. P. Le, S. R. Sanders, E. Alon, "Design techniques for fully integrated switched-capacitor DC-DC converters," *IEEE Journal of Solid State Circuits*, Vol. 46, No.9, pp. 2120–2131, Sept. 2011.

[46] S. Akhtar, P. Litmanen, M. Ipek, J. H. C. Lin, S. Pennisi, F. J. Huang, and R. B. Staszewski, "Analog path for triple band WCDMA polar modulated transmitter in 90nm CMOS," *Proc. of 2007 IEEE Radio Frequency Integrated Circuits Symp.*, June 2007, pp. 185–188.

[47] J. Shu, M. Cai, "A low supply-dependence fully-MOSFET voltage reference for low-voltage and low-power," *IEEE Asia Pacific Conference on Circuits and Systems*, Dec. 2008, pp. 662–665.

[48] I. Doms, P. Merken, C. van Hoof, R. P. Mertens, "Capacitive power management circuit for micropower thermoelectric generators with a 1.4μA controller", *IEEE Journal of Solid State Circuits*, Vol. 44, No. 10, pp. 2824–2833, Oct. 2009.

[49] J. Kim, J. Kim, C. Kim, "A regulated charge pump with a low-power integrated optimum power point tracking algorithm for indoor solar

energy harvesting," *IEEE Transaction on Circuits and Systems II*, Vol. 58, No. 12, Dec. 2011.

[50] M. El-nozahi, A. Amer, J. Torres, K. Entesari, E. Sanchez-Sinencio, "A 25mA 0.13μm CMOS LDO regulator with power supply rejection better than −56dB up to 10MHz using feedforward ripple cancellation technique," *IEEE International Solid-State Circuit Conference*, Feb. 2009, pp. 330–331.

[51] J. J. Chen, M. S. Lin, H. C. Lin, Y. S. Hwang, "Sub-1V capacitor-free low-power-consumption LDO with digital controlled loop," *IEEE Asia Pacific conference on circuits and systems*, Dec. 2008, pp. 526–529.

[52] M. Al-Shyoukh, H. Lee, R. Perez, "A transient-enhanced low-quiescent current low-dropout regulator with buffer impedance attenuation," *IEEE Journal of Solid State Circuits*, Vol. 42, No.8, pp. 1732–1742, Aug. 2007.

[53] M. D. Seeman, S. R. Sanders, J. M. Rabaey, "An ultra-low-power power management IC for energy-scavenged wireless sensor nodes," *IEEE 39th Power Electronics Specialists Conference Proceedings, PESC*, June 2008, pp. 925–931.

Index

About the Authors

Nitz Saputra received the B.Eng. degree with honors from the Nanyang Technological University, Singapore in 2002, and the M.Sc. (cum laude), and Ph.D. degrees from Delft University of Technology, the Netherlands, in 2005 and 2012 respectively. He worked in industry with Marvell Asia, Singapore (2002–2003), DIMES, the Netherlands (2006–2007), and Broadcom Netherlands B.V., the Netherlands (2012–2013). Since October 2013 he joined Qualcomm Inc., USA. His current research interests include low-power analog, mixed-signal, and RF circuit design in CMOS and advanced FinFET technology.

John R. Long received the B.Sc. in Electrical Engineering from the University of Calgary in 1984, and the M.Eng. and Ph.D. degrees in Electronics from Carleton University in Ottawa, Canada, in 1992 and 1996, respectively. He worked in industry for 12 years in the Advanced Technology Laboratory at Bell-Northern Research in Ottawa, and an academic at the University of Toronto (1996–2002) and as Chair of the Electronics Research Laboratory at the Delft University of Technology in the Netherlands (2002–2014). In January 2015 he was appointed Professor in Electrical and Computer Engineering at the University of Waterloo in Canada. His current research interests include low-power and broadband circuits for highly-integrated wireless transceivers, energy-efficient wireless sensors, mm-wave IC design, and electronics design for high-speed data communications.

Professor Long is editor-in-chief of the IEEE RFIC Virtual Journal, and a Distinguished Lecturer for the IEEE Solid-State Circuits Society. He has served as an organizer and program committee member for many IEEE-sponsored conferences, including: the ISSCC (1998–2011), the IEEE-BCTM (2006 conference General Chair and local organizer), the ESSCIRC (2003–2014) and EuMIC (conference co-Chair in 2009 and 2012). Professor Long is also a former Associate Editor of the IEEE Journal of Solid-State Circuits.

He received the NSERC Doctoral Prize, Douglas R. Colton and Governor General's Medals for research excellence, and is co-recipient of Best Paper Awards from ISSCC (2000 and 2007), BCTM (2003 and 2014), the RFIC Symposium (2006, 2011 and 2013), and EuMW in 2006.

CPSIA information can be obtained
at www.ICGtesting.com
Printed in the USA
BVOW11*1628220217
476413BV00001B/1/P

9 788793 519169